Health Care Systems

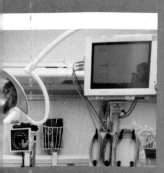
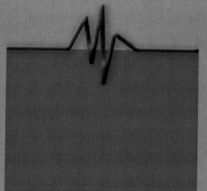
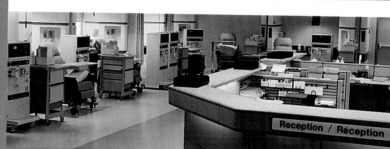

Properly Applying Codes and Standards

Contents

Contents

Acknowledgements

Acknowledgements

Hubbell Wiring Device-Kellems
Jonathan Gilbert
Lawson Electric Co.
Lutron Electronics Co., Inc.
Michael J. Johnston
Pass & Seymour/Legrand
Stephen M. Lipster

QR Codes

Atkore International, Inc.
Judene Bartley, ECSI Inc.
Caterpillar
COOPER Bussmann
COOPER Safety
Emerson Electric Co.
General Electric Company
Pass & Seymour/Legrand
Schneider Electric
Southwire Company
U.S. Department of Labor
    Occupational Safety and Health
    Administration

NFPA 70®, *National Electrical Code* and *NEC*® are registered trademarks of the National Fire Protection Association, Quincy, MA.

This material is continually reviewed and evaluated by Training Directors who are also members of the NJATC Inside Education Committee. The invaluable input provided by these individuals allows for the development of instructional material that is of the absolute highest quality. At the time of this printing, the Inside Education Committee was composed of the following members:

Kathleen Barber, San Carlos, CA
Byron Benton, San Leandro, CA
John Biondi, Vineland, NJ
Richard M. Brooks, Dayton, OH
Eric S. Davis, Warren, OH
Lawrence Hidalgo, Lansing, MI
Gregory A. Hojdila, Beaver, PA
Kenneth Jania, Merrillville, IN

Tony Lewis, Tacoma, WA
Dave McCraw, Tampa, FL
David Milazzo, Paramus, NJ
Tom Minder, Fairbanks, AK
Tony Naylor, Wichita, KS
Janet Skipper, Winter Park, FL
Jim Sullivan, Winter Park, FL
Dennis Williamson, Kennewick, WA

Features

Blue **Headers** and **Subheaders** organize information within the text.

Facts offer additional information related to Health Care Systems.

Figures, including photographs and artwork, clearly illustrate concepts from the text.

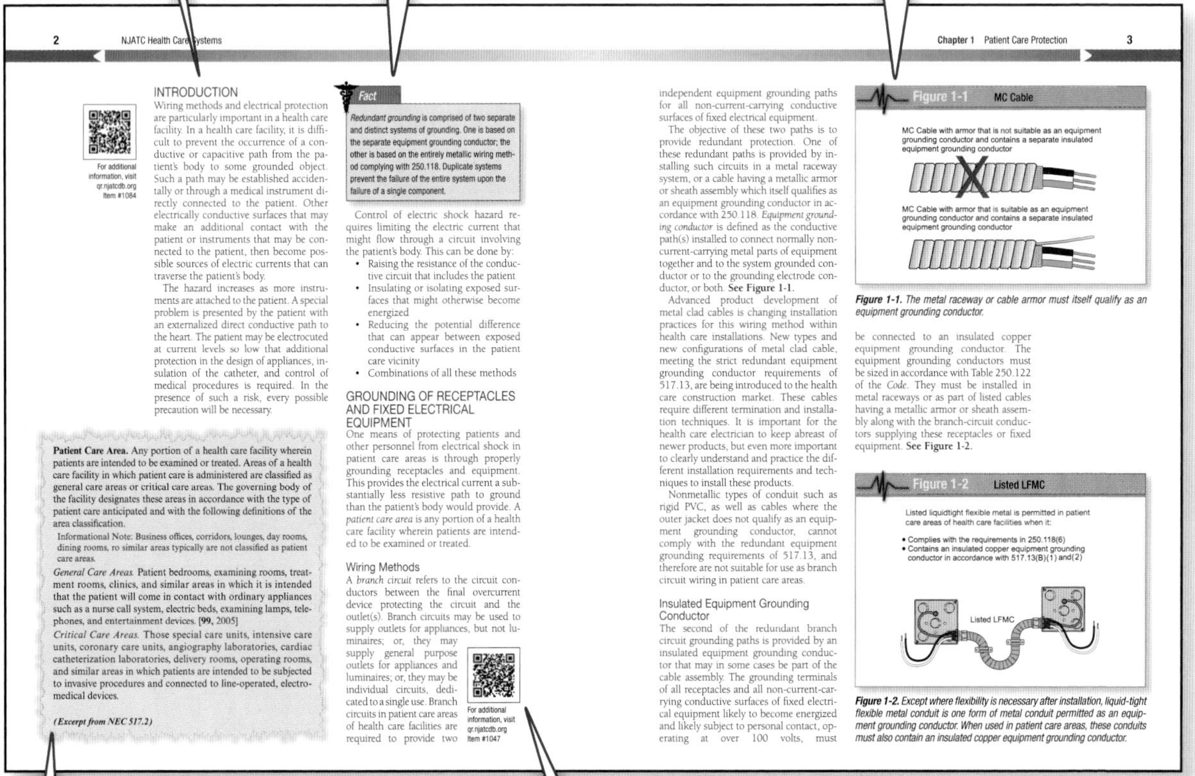
2 NJATC Health Care Systems

INTRODUCTION

Wiring methods and electrical protection are particularly important in a health care facility. In a health care facility, it is difficult to prevent the occurrence of a conductive or capacitive path from the patient's body to some grounded object. Such a path may be established accidentally or through a medical instrument directly connected to the patient. Other electrically conductive surfaces that may make an additional contact with the patient or instruments that may be connected to the patient, then become possible sources of electric currents that can traverse the patient's body.

The hazard increases as more instruments are attached to the patient. A special problem is presented by the patient with an externalized direct conductive path to the heart. The patient may be electrocuted at current levels so low that additional protection in the design of appliances, insulation of the catheter, and control of medical procedures is required. In the presence of such a risk, every possible precaution will be necessary.

For additional information, visit qr.njatcdb.org Item #1084

Patient Care Area. Any portion of a health care facility wherein patients are intended to be examined or treated. Areas of a health care facility in which patient care is administered are classified as general care areas or critical care areas. The governing body of the facility designates these areas in accordance with the type of patient care anticipated and with the following definitions of the area classification.

Informational Note: Business offices, corridors, lounges, day rooms, dining rooms, ro similar areas typically are not classified as patient care areas.

General Care Areas. Patient bedrooms, examining rooms, treatment rooms, clinics, and similar areas in which it is intended that the patient will come in contact with ordinary appliances such as a nurse call system, electric beds, examining lamps, telephones, and entertainment devices. [99, 2005]

Critical Care Areas. Those special care units, intensive care units, coronary care units, angiography laboratories, cardiac catheterization laboratories, delivery rooms, operating rooms, and similar areas in which patients are intended to be subjected to invasive procedures and connected to line-operated, electromedical devices.

(Excerpt from NEC 517.2)

Fact

Redundant grounding is comprised of two separate and distinct systems of grounding. One is based on the separate equipment grounding conductor; the other is based on the entirely metallic wiring method complying with 250.118. Duplicate systems prevent the failure of the entire system upon the failure of a single component.

Control of electric shock hazard requires limiting the electric current that might flow through a circuit involving the patient's body. This can be done by:

- Raising the resistance of the conductive circuit that includes the patient
- Insulating or isolating exposed surfaces that might otherwise become energized
- Reducing the potential difference that can appear between exposed conductive surfaces in the patient care vicinity
- Combinations of all these methods

GROUNDING OF RECEPTACLES AND FIXED ELECTRICAL EQUIPMENT

One means of protecting patients and other personnel from electrical shock in patient care areas is through properly grounding receptacles and equipment. This provides the electrical current a substantially less resistive path to ground than the patient's body would provide. A *patient care area* is any portion of a health care facility wherein patients are intended to be examined or treated.

Wiring Methods

A *branch circuit* refers to the circuit conductors between the final overcurrent device protecting the circuit and the outlet(s). Branch circuits may be used to supply outlets for appliances, but not luminaires; or, they may supply general purpose outlets for appliances and luminaires; or, they may be individual circuits, dedicated to a single use. Branch circuits in patient care areas of health care facilities are required to provide two

For additional information, visit qr.njatcdb.org Item #1047

Chapter 1 Patient Care Protection 3

independent equipment grounding paths for all non-current-carrying conductive surfaces of fixed electrical equipment.

The objective of these two paths is to provide redundant protection. One of these redundant paths is provided by installing such circuits in a metal raceway system, or a cable having a metallic armor or sheath assembly which itself qualifies as an equipment grounding conductor in accordance with 250.118. *Equipment grounding conductor* is defined as the conductive path(s) installed to connect normally non-current-carrying metal parts of equipment together and to the system grounded conductor or to the grounding electrode conductor, or both. **See Figure 1-1.**

Advanced product development of metal clad cables is changing installation practices for this wiring method within health care installations. New types and new configurations of metal clad cable, meeting the strict redundant equipment grounding conductor requirements of 517.13, are being introduced to the health care construction market. These cables require different termination and installation techniques. It is important for the health care electrician to keep abreast of newer products, but even more important to clearly understand and practice the different installation requirements and techniques to install these products.

Nonmetallic types of conduit such as rigid PVC, as well as cables where the outer jacket does not qualify as an equipment grounding conductor, cannot comply with the redundant equipment grounding requirements of 517.13, and therefore are not suitable for use as branch circuit wiring in patient care areas.

Insulated Equipment Grounding Conductor

The second of the redundant branch circuit grounding paths is provided by an insulated equipment grounding conductor that may in some cases be part of the cable assembly. The grounding terminals of all receptacles and all non-current-carrying conductive surfaces of fixed electrical equipment likely to become energized and likely subject to personal contact, operating at over 100 volts, must be connected to an insulated copper equipment grounding conductor. The equipment grounding conductors must be sized in accordance with Table 250.122 of the *Code*. They must be installed in metal raceways or as part of listed cables having a metallic armor or sheath assembly along with the branch-circuit conductors supplying these receptacles or fixed equipment. **See Figure 1-2.**

Figure 1-1 MC Cable

MC Cable with armor that is not suitable as an equipment grounding conductor and contains a separate insulated equipment grounding conductor

MC Cable with armor that is suitable as an equipment grounding conductor and contains a separate insulated equipment grounding conductor

Figure 1-1. The metal raceway or cable armor must itself qualify as an equipment grounding conductor.

Figure 1-2 Listed LFMC

Listed liquidtight flexible metal is permitted in patient care areas of health care facilities when it:

- Complies with the requirements in 250.118(6)
- Contains an insulated copper equipment grounding conductor in accordance with 517.13(B)(1) and (2)

Listed LFMC

Figure 1-2. Except where flexibility is necessary after installation, liquid-tight flexible metal conduit is one form of metal conduit permitted as an equipment grounding conductor. When used in patient care areas, these conduits must also contain an insulated copper equipment grounding conductor.

Code Excerpts are "ripped" from the *NEC* or other sources.

For additional information related to QR Codes, visit qr.njatcdb.org Item #1079

Quick Response Codes (QR Codes) create a link between the textbook and the Internet. They can be scanned using Smartphone applications to obtain additional information online. (To access the information without using a Smartphone, visit qr.njatc.org and enter the referenced Item #.)

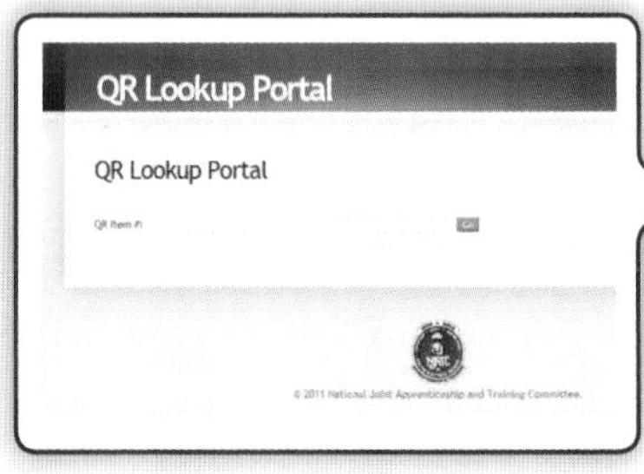

Features

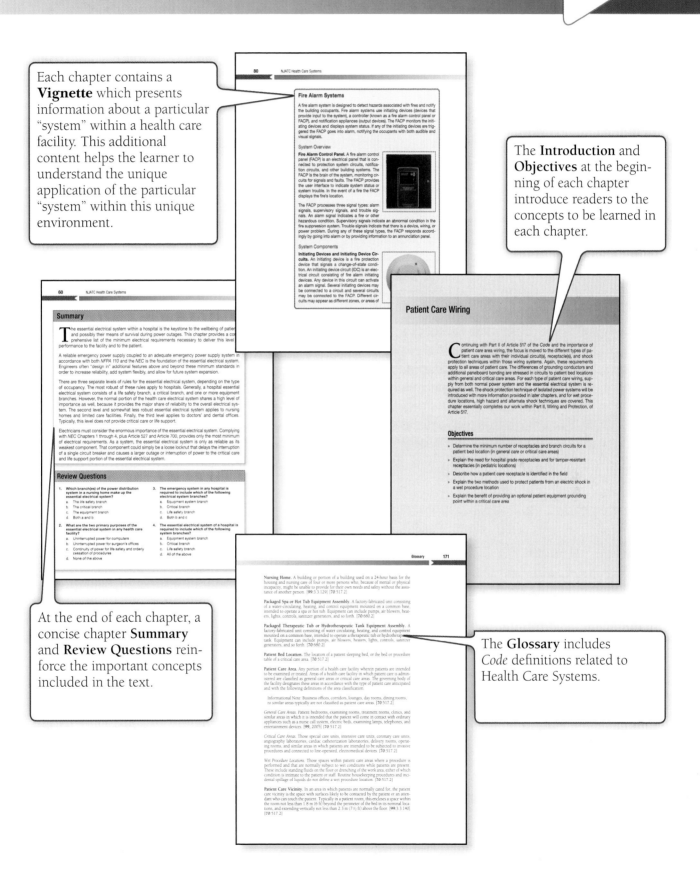

Each chapter contains a **Vignette** which presents information about a particular "system" within a health care facility. This additional content helps the learner to understand the unique application of the particular "system" within this unique environment.

The **Introduction** and **Objectives** at the beginning of each chapter introduce readers to the concepts to be learned in each chapter.

At the end of each chapter, a concise chapter **Summary** and **Review Questions** reinforce the important concepts included in the text.

The **Glossary** includes *Code* definitions related to Health Care Systems.

About this Book

Health care facilities present special challenges for the contractor, designer, engineer, electrical inspector, electrical worker, and health care facility personnel. A thorough understanding of the general installation methods and materials and associated NFPA 70: *National Electrical Code® (NEC)* requirements is imperative. Those involved in the design, installation, and maintenance of health care electrical systems must also have a thorough understanding of the unique provisions and considerations of health care buildings and occupancies. *Health Care Systems* is a specialized *Code*-based textbook and an invaluable resource to help those involved with installing health care facility electrical systems gain an understanding of a number of the concepts and applicable requirements related to health care facilities.

Health Care Systems introduces and explores a number of codes and standards applicable to health care installations, in addition to those contained in the *NEC®*. These include, but are not limited to, NFPA 99, *Standard for Health Care Facilities*; NFPA 72, *National Fire Alarm and Signaling Code*; NFPA 101, *Life Safety Code®*; and NFPA 110, *Standard for Emergency and Standby Power Systems*. This 2nd edition textbook contains 8 chapters, six of which focus directly on Article 517 of the 2011 *NEC*. The textbook addresses topics such as patient care protection and wiring, essential electrical systems, and isolated power systems. New for this edition is the content of Chapter 8, *Working in Operational Facilities*. This chapter introduces important safety and planning concepts within an operational health care facility.

Introduction

Health care facilities present special challenges for the electrical worker, both from a technical perspective and from a personnel one. For instance, health care facilities have special requirements for redundant power supplies. This results in numerous technical issues, ranging from the installation of backup generators to the type of wiring device to use. In addition to the technical issues, there are the human ones. Work needs to be done differently to avoid disturbing patients. There are also special PPE requirements to protect the facility from contamination and to protect the worker from infection.

ASSUMPTIONS

This book is based on the requirements contained in Articles 517 and 680 of NFPA 70: *National Electrical Code®* (*NEC* or *Code*). Article 517 Health Care Facilities is specifically dedicated to electrical work in health care facilities. By reference, it includes portions of Article 680 Swimming Pools, Fountains, and Similar Installations related to water therapy.

This is a book for the experienced electrical worker who wants to prepare for work in a health care setting. The assumption is that an experienced person is already familiar with Chapters 1-4 of the *Code*. Therefore, in treating the materials from Chapter 5 Article 517, and Chapter 6 Article 680, no attempt will be made to explain references to the earlier chapters of the *Code*. For instance, numerous references to bonding, grounding, the use of ground-fault circuit interrupters, and the dangers of shock occur throughout this book. These, however, are all topics that should be familiar to the reader, and therefore they are not explained unless there is some unique aspect that applies to them in the health care context.

BOOK OVERVIEW

The book is organized as follows:
- Introduction
- Patient Care Protection
- Patient Care Wiring
- Essential Electrical Systems
- Inhalation Anesthetizing Locations
- Diagnostic Imaging Equipment Installations
- Isolated Power Systems
- Pools and Tubs for Therapeutic Use
- Working in Operational Facilities
- Glossary

Article 517 also references Article 700, Emergency Systems, and *NFPA 72, National Fire Alarm Code*. This book does not attempt to cover these topics, however, because they are not specifically directed at health care facilities. Clearly, a hospital must have a system designed for emergency power and a robust fire alarm system. However, other buildings require emergency systems and fire alarms, and there are other books devoted to these topics. This book only touches on them insofar as they may need to be adapted or modified for use in a health care facility; this is how Article 517 treats them.

HISTORY OF ARTICLE 517

The *NEC* is a publication of the National Fire Protection Association (NFPA) and is also referenced as *NFPA 70*. The *NEC* was first published in 1897. It is updated every three years, the most recent being the 2011 edition (updated in 2010). The *NEC* contains frequent references to other publications of the NFPA. Wherever material in the *NEC* has been extracted from other publications of the NFPA, those other publications are referenced in brackets by NFPA number. For instance, the definition for *electrical life-support equipment* appears in the *NEC* as follows:

Electrical Life-Support Equipment. Electrically powered equipment whose continuous operation is necessary to maintain a patient's life. [99:3.3.37]

Many of the requirements contained in Article 517 of the *Code* come from *NFPA 99, Standard for Health Care Facilities*. *NFPA 99* was first issued in 1984 as a compilation of 12 previous NFPA documents dealing with health care facilities, all of which fell under the responsibility of the NFPA Health Care Facilities Technical Correlating Committee. The Health Care Facilities technical committees are responsible for developing and maintaining criteria for performance, maintenance, and testing of electrical systems and appliances. Code-Making Panel 15 (CMP-15) is the current electrical code technical committee responsible for electrical system installation rules in the *NEC*.

This book uses the *NEC* as its primary reference and will not attempt to provide duplicate references for every item. If the *NEC* already contains a cross reference to another NFPA document, this book will only reference the *NEC* entry. However, in cases where even though the *NEC* references another NFPA publication, the full force of the applicable code is only understood by returning to the original, this book will reference both the *NEC* and, as appropriate, other NFPA documents. Special note should be made of *NFPA 99-2005: Standard for Health Care Facilities* and of *NFPA 101-2006: Life Safety Code*. Both of these are referenced frequently by *NEC* Article 517. Another significant NFPA document dealing with the essential electrical systems for hospitals is *NFPA 110-2010, Standard for Emergency and Standby Power Systems*.

STANDARDS APPLICATION

As is consistent with most other safety codes and standards, the standards of the *Code* and *NFPA 99* are not generally intended to be applied retroactively. They are applicable to new construction and equipment. Each existing facility should be considered individually when applying requirements from the *Code*. NFPA **99**:1.3.2 provides a general rule indicating that this standard applies only to new construction and new equipment in health care facilities unless a specific modification within an individual chapter of the standard applies. If a health care facility is altered, renovated, or modernized, only that portion of the facility or directly affected portions of the systems would be required to meet the requirements in *NFPA 99*. Where a renovation, alteration, or modernization of a health care facility adversely impacts existing systems or performance requirements of systems, additional upgrades are necessary and required (Richard P. Bielen, *Health Care Facilities Handbook*, 8th ed., NFPA 2005).

ENGINEERING DESIGNS AND SPECIFICATIONS

It is important to be familiar with all applicable general *NEC* requirements in Chapters 1 through 4, as these apply to all facilities, including health care facilities. It is also important to become familiar with the special requirements of Chapters 5, 6, and 7, which will be covered in more detail in this text. According to Section 90.3, Chapters 1, 2, 3 and 4 apply generally. Chapters 5, 6, and 7 supplement or modify the requirements of the first four chapters.

NEC Figure 90.3 Code Arrangement

PLAN
BUILD
USE

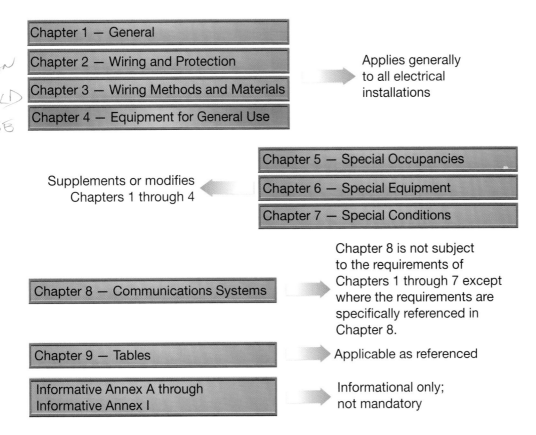

Chapter 1 — General
Chapter 2 — Wiring and Protection
Chapter 3 — Wiring Methods and Materials
Chapter 4 — Equipment for General Use

Applies generally to all electrical installations

Chapter 5 — Special Occupancies
Chapter 6 — Special Equipment
Chapter 7 — Special Conditions

Supplements or modifies Chapters 1 through 4

Chapter 8 — Communications Systems

Chapter 8 is not subject to the requirements of Chapters 1 through 7 except where the requirements are specifically referenced in Chapter 8.

Chapter 9 — Tables

Applicable as referenced

Informative Annex A through Informative Annex I

Informational only; not mandatory

Another important aspect of electrical installations in health care facilities usually involves an electrical design by professional engineers and designers. This is especially true for hospital electrical systems. While the *NEC*, *NFPA 99*, and other standards provide the minimum conditions for electrical designs in these facilities, engineers and designers often exceed the minimums in the *Code* and other standards as a part of the individual project plans and specifications. These vital construction documents must be followed. A good working relationship should be established between the designing professional and the installing contractor for optimum results. Another consideration is that many of the administrative rules and regulations of the applicable authority having jurisdiction (AHJ) require engineered plans and specifications to be followed exactly, unless alternatives or deviations are otherwise addressed or under the consent of the responsible engineering firm. This is important when considering alternative methods in the electrical wiring system for application in any health care facility. Before working on any electrical system in a health care facility, one should always check with the authority having jurisdiction for any other local electrical code rules that may be applied to the installation that are in addition to the *NEC* requirements. Communication and cooperation between the design team and installing contractor are essential parts of successful electrical installations in any occupancy, but especially in health care facilities.

Patient Care Protection

General requirements for wiring and protection (of equipment and personnel) in health care facilities are covered in Part II of Article 517 of the *Code*. These requirements apply to patient care areas in all health care facilities. They do not cover business offices, corridors, waiting rooms, and the like; nor do they cover areas of nursing homes or limited care facilities where rooms are used exclusively as sleeping areas. This chapter and the next chapter of this textbook reinforce each other. They should be read carefully and consecutively. They both deal with what the *Code* calls General Installation — Construction Criteria. Most often omitted from reading, Section 517.11 and its Informational Note set down the fundamental issues of electrical construction that will minimize electrical hazards within patient care areas of all health care facilities. These performance requirements and related information must always be the number one goal of electrical construction and maintenance within patient care areas. Chapter 1 deals with these general requirements for patient care protection from a circuit grounding and ground fault protection (for feeders) point of view.

Objectives

» Describe the unique shock and electrocution hazards that exist for patients and personnel in health care facilities

» Determine which wiring methods are suitable for branch circuits and feeders that supply patient care areas

» Explain the reasons for redundant equipment grounding in branch circuits supplying patient care areas and the rules that apply to bonding of panelboards that serve the same patient care vicinity

» Determine the use and different types of ground-fault protection systems for feeders in a health care facility

» Describe the term selectivity as applied to the levels of ground-fault protection systems for feeders in a health care facility

Chapter 1

Table of Contents

INTRODUCTION

Wiring methods and electrical protection are particularly important in a health care facility. In a health care facility, it is difficult to prevent the occurrence of a conductive or capacitive path from the patient's body to some grounded object. Such a path may be established accidentally or through a medical instrument directly connected to the patient. Other electrically conductive surfaces that may make an additional contact with the patient or instruments that may be connected to the patient, then become possible sources of electric currents that can traverse the patient's body.

The hazard increases as more instruments are attached to the patient. A special problem is presented by the patient with an externalized direct conductive path to the heart. The patient may be electrocuted at current levels so low that additional protection in the design of appliances, insulation of the catheter, and control of medical procedures is required. In the presence of such a risk, every possible precaution will be necessary.

For additional information, visit qr.njatcdb.org Item #1084

Patient Care Area. Any portion of a health care facility wherein patients are intended to be examined or treated. Areas of a health care facility in which patient care is administered are classified as general care areas or critical care areas. The governing body of the facility designates these areas in accordance with the type of patient care anticipated and with the following definitions of the area classification.

Informational Note: Business offices, corridors, lounges, day rooms, dining rooms, ro similar areas typically are not classified as patient care areas.

General Care Areas. Patient bedrooms, examining rooms, treatment rooms, clinics, and similar areas in which it is intended that the patient will come in contact with ordinary appliances such as a nurse call system, electric beds, examining lamps, telephones, and entertainment devices. [99, 2005]

Critical Care Areas. Those special care units, intensive care units, coronary care units, angiography laboratories, cardiac catheterization laboratories, delivery rooms, operating rooms, and similar areas in which patients are intended to be subjected to invasive procedures and connected to line-operated, electro-medical devices.

(Excerpt from NEC 517.2)

Fact

Redundant grounding is comprised of two separate and distinct systems of grounding. One is based on the separate equipment grounding conductor; the other is based on the entirely metallic wiring method complying with 250.118. Duplicate systems prevent the failure of the entire system upon the failure of a single component.

Control of electric shock hazard requires limiting the electric current that might flow through a circuit involving the patient's body. This can be done by:

- Raising the resistance of the conductive circuit that includes the patient
- Insulating or isolating exposed surfaces that might otherwise become energized
- Reducing the potential difference that can appear between exposed conductive surfaces in the patient care vicinity
- Combinations of all these methods

GROUNDING OF RECEPTACLES AND FIXED ELECTRICAL EQUIPMENT

One means of protecting patients and other personnel from electrical shock in patient care areas is through properly grounding receptacles and equipment. This provides the electrical current a substantially less resistive path to ground than the patient's body would provide. A *patient care area* is any portion of a health care facility wherein patients are intended to be examined or treated.

Wiring Methods

A *branch circuit* refers to the circuit conductors between the final overcurrent device protecting the circuit and the outlet(s). Branch circuits may be used to supply outlets for appliances, but not luminaires; or, they may supply general purpose outlets for appliances and luminaires; or, they may be individual circuits, dedicated to a single use. Branch circuits in patient care areas of health care facilities are required to provide two

For additional information, visit qr.njatcdb.org Item #1047

independent equipment grounding paths for all non-current-carrying conductive surfaces of fixed electrical equipment.

The objective of these two paths is to provide redundant protection. One of these redundant paths is provided by installing such circuits in a metal raceway system, or a cable having a metallic armor or sheath assembly which itself qualifies as an equipment grounding conductor in accordance with 250.118. *Equipment grounding conductor* is defined as the conductive path(s) installed to connect normally non-current-carrying metal parts of equipment together and to the system grounded conductor or to the grounding electrode conductor, or both. **See Figure 1-1.**

Advanced product development of metal clad cables is changing installation practices for this wiring method within health care installations. New types and new configurations of metal clad cable, meeting the strict redundant equipment grounding conductor requirements of 517.13, are being introduced to the health care construction market. These cables require different termination and installation techniques. It is important for the health care electrician to keep abreast of newer products, but even more important to clearly understand and practice the different installation requirements and techniques to install these products.

Nonmetallic types of conduit such as rigid PVC, as well as cables where the outer jacket does not qualify as an equipment grounding conductor, cannot comply with the redundant equipment grounding requirements of 517.13, and therefore are not suitable for use as branch circuit wiring in patient care areas.

Insulated Equipment Grounding Conductor

The second of the redundant branch circuit grounding paths is provided by an insulated equipment grounding conductor that may in some cases be part of the cable assembly. The grounding terminals of all receptacles and all non-current-carrying conductive surfaces of fixed electrical equipment likely to become energized and likely subject to personal contact, operating at over 100 volts, must

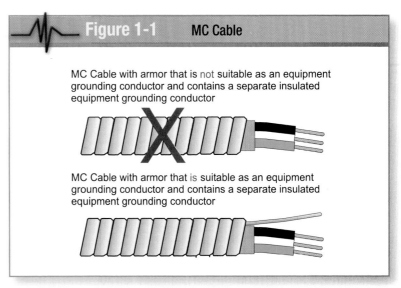

Figure 1-1. *The metal raceway or cable armor must itself qualify as an equipment grounding conductor.*

be connected to an insulated copper equipment grounding conductor. The equipment grounding conductors must be sized in accordance with Table 250.122 of the *Code*. They must be installed in metal raceways or as part of listed cables having a metallic armor or sheath assembly along with the branch-circuit conductors supplying these receptacles or fixed equipment. **See Figure 1-2.**

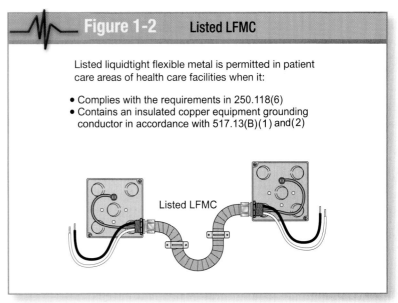

Figure 1-2. *Except where flexibility is necessary after installation, liquid-tight flexible metal conduit is one form of metal conduit permitted as an equipment grounding conductor. When used in patient care areas, these conduits must also contain an insulated copper equipment grounding conductor.*

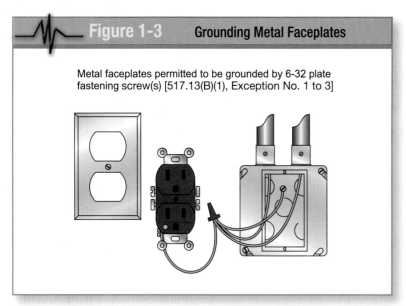

Figure 1-3. Metal faceplates are permitted to be grounded by 6-32 plate fastening screw(s).

The only exceptions to these rules are as follows:

1. Metal faceplates may be connected to the equipment grounding conductor by means of a metal mounting screw(s) securing the faceplate to a grounded outlet box or grounded wiring device. **See Figure 1-3.**
2. Luminaires more than 7¹/₂ feet (2.3 m) above the floor and switches located outside of the patient care vicinity may be connected to an

Figure 1-4   Patient Care Area Lighting

Luminaires above 7½ ft do not require insulated ground

Luminaires

Patient care area

8 ft.

Switches located outside patient vicinity are not required to be grounded by insulated grounding conductor.

Figure 1-4. Luminaires above 7¹/₂ feet are required to be grounded by at least one equipment grounding return path complying with 517.13(A).

equipment grounding return path that complies with the requirements above under "Wiring Methods." **See Figure 1-4.**

In an area in which patients are normally cared for, the *patient care vicinity* is the space with surfaces likely to be contacted by the patient or an attendant who can touch the patient. Typically in a patient room, this encloses a space within the room not less than 1.8 m (6 ft) beyond the perimeter of the bed in its nominal locations, and extending vertically not less than 2.3 m (7¹/₂ ft) above the floor. [**99**:3.3.140]

Grounding System Testing

NFPA 99, Standard for Health Care Facilities requires that the integrity of the grounding path provided by the wiring method in patient care areas of health care facilities be tested before acceptance of the initial installation and after any alterations or replacement of the electrical system is made. [**99**:4.3.3.1.1] Both voltage and impedance measurements must be made of exposed conductive surfaces including the grounding contacts of receptacles in the patient-care vicinity. Excluded from the testing requirements are small, wall-mounted conductive surfaces not likely to become energized, such as surface-mounted towel and soap dispensers, mirrors, and the like. Also exempt are large metal conductive surfaces not likely to become energized, such as window frames, door frames, and drains. The grounding system, including both metallic raceway and equipment grounding conductor, is to be tested as an integral system. Removing equipment grounds from receptacles or equipment is not required or recommended.

NFPA 99 requires that voltage and impedance measurements be measured against a reference point. The reference point is permitted to be one of the following:

1. A reference grounding point
   (A *reference grounding point* is the ground bus of the panelboard or isolated power system panel supplying the patient care area.)

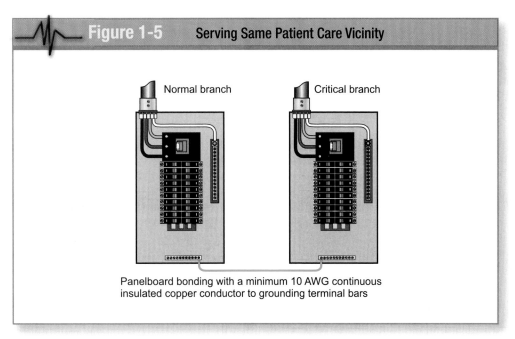

Figure 1-5   **Serving Same Patient Care Vicinity**

Normal branch   Critical branch

Panelboard bonding with a minimum 10 AWG continuous
insulated copper conductor to grounding terminal bars

Figure 1-5. *Panelboards are bonded by connecting the equipment grounding terminal bars in accordance with 517.14.*

2. A grounding point in the room under test that is electrically remote from the equipment under test, such as a metal water pipe
3. The grounding contact of a receptacle that is powered from a different branch circuit from the receptacle being tested [**99**:4.3.3.1.2]

The maximum acceptable values for new construction acceptance testing that must be met are as follows:
1. Voltage limit is 20 mV
2. Ohm impedance limit is 0.1 ohm
3. For isolated ground circuits (quiet ground systems), the impedance limit is 0.2 ohm

In wet procedure locations, fixed receptacles, equipment connected by cord and plug, and fixed electrical equipment must be tested:
1. When first installed
2. Where there is evidence of damage
3. After any repairs
4. At intervals not exceeding 6 months

4.3.4.1 of *NFPA 99* provides requirements for the testing interval for hospital grade receptacles at patient bed locations and in anesthetizing locations. A *patient bed location* is the location of a patient sleeping bed, or the bed or procedure

table of a critical care area.

The testing interval for patient bed locations and anesthetizing locations should be as follows:
1. Testing be performed after initial installation, replacement, or servicing of the device
2. Additional testing be performed at intervals defined by documented performance data

PANELBOARD BONDING

The equipment grounding terminal buses of the normal and essential branch-circuit panelboards serving the same individual patient care vicinity must be connected together with an insulated continuous copper conductor not smaller than 10 AWG. Where two or more panelboards serving the same individual patient care vicinity are served from separate transfer switches on the emergency system, the equipment grounding terminal buses of those panelboards must be connected together with an insulated continuous copper conductor not smaller than 10 AWG. **See Figure 1-5.**

Note the terms "normal," "essential," and "emergency" in the above paragraph. These are terms with specific meanings, explained

---

Fact

The terms *patient bed location*, *patient care area*, and *patient care vicinity* all have technical distinctions that should be reviewed before working in a health care facility. They are not always the same thing. They are defined in this chapter and again in the glossary at the end of this book.

in more detail in the following chapter on the "essential electrical system." One should plan on revisiting both this chapter and the next frequently to understand how their topics depend upon each other.

This conductor may be broken in order to terminate on the equipment grounding terminal bus in each panelboard.

Bonding all equipment grounding conductor terminal bars of all panelboards serving the same patient vicinity helps minimize the possible potential differences. The *Code* only requires a 10 AWG connection, but the connection may be larger. For example, if the panelboards were located far enough apart, voltage drop could become a concern.

RECEPTACLES WITH INSULATED GROUNDING TERMINALS

For the 2011 edition of the *NEC*, Section 517.16 has been revised to prohibit the use of receptacles with insulating grounding terminals, that is, isolated grounding receptacles, in patient care areas of all health care facilities. However, outside of patient care areas, these circuits and devices continue to be permitted.

Provided the equipment is not directly related to patient care, or located within

Fact

The *Code* change substantiation for no longer permitting isolated grounding receptacles in patient care areas is because "… the reduction of electrical noise should not be taken more seriously than protecting the patient, particularly in an invasive procedure area."

patient care areas, medical equipment manufacturers may continue to specify isolated ground receptacles in order to reduce the amount of electrical magnetic interference (EMI) in the grounding circuit for sensitive electronics. In these special circuits, an additional isolated, insulated equipment grounding conductor must be installed with the circuit conductors. This is in addition to the required equipment grounding conductor.

Receptacles with insulated grounding terminals must be identified and the identification must be visible after installation. The requirements for isolated grounding receptacles are found in 250.146(D). **See Figure 1-6.**

Caution is important in specifying such a system with receptacles having insulated grounding terminals. The grounding impedance is controlled only by the equipment grounding conductors and does not benefit functionally from any parallel grounding paths. This type of installation is typically used where a reduction of electrical noise (electromagnetic interference) is necessary and parallel grounding paths are to be avoided. Since parallel equipment grounding paths are imperative in branch circuits within patient care areas, it is understandable that these isolated grounding receptacles and their circuit are no longer permitted in patient care areas within all health care facilities.

GROUND-FAULT PROTECTION

An electric arc, which generates a tremendous amount of heat, is readily maintained at 277 volts to ground. This is not a solidly grounded (bolted) connection, and the current is limited by the impedance of the short circuit; therefore, not

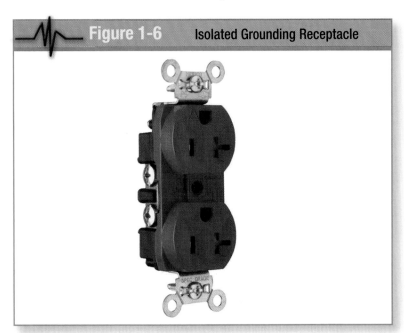

Figure 1-6   **Isolated Grounding Receptacle**

Figure 1-6. *This isolated grounding receptacle has an identifying mark on its face. New for the 2011 NEC, Section 517.16 no longer allows the use of isolated grounding receptacles in patient care areas.*

enough current will flow in the circuit to open the overcurrent device ahead of the fault. A great deal of damage can be done to the electrical equipment while the arc is burning.

Ground-fault protection must be carefully designed for health care facilities, especially for those areas of hospitals or other buildings with life safety feeders and critical care feeders to areas that use electrical life support equipment. A *feeder* includes all circuit conductors between the service equipment, the source of a separately derived system, or other power supply source and the final branch-circuit overcurrent device.

Types of Ground-Fault Protection Equipment

There are two basic types of ground-fault protection equipment:

1. Residual, which is sometimes referred to as ground-strap or ground-return equipment
2. Zero-sequence, which has multiple forms

Both types of GFP equipment are designed to protect downstream equipment from destructive high impedance faults. This equipment will not protect equipment or the system on its line side from line-to-ground faults because this fault current will not pass through and be detected by the ground-fault sensing equipment.

Residual-Type (Neutral Ground Strap) System

The neutral ground-strap type of ground-fault protection consists of a current transformer, control relay, and usually a shunt trip circuit breaker. The main bonding jumper passes through the current transformer. **See Figure 1-7.**

This type of equipment is the least expensive, but it is limited to application at the main service. Neutral ground strap equipment cannot be used on the load side of the main bonding jumper at the service or the load side of a system bonding jumper in a separately derived system. The neutral ground strap system functions by sensing ground-fault current through the main bonding jumper during a ground

fault. Although there are many parallel circuits for ground-fault current, the relative impedance of the main circuit is so low compared to all the parallel bypass circuits that a large percentage of the total ground-fault current will be present in the main bonding jumper during a ground fault. This fault current is then sensed by the current transformer. If a current transformer is placed about the main bonding jumper and has as its secondary feeding an adjustable magnetic current relay, the relay may be adjusted to close a contact at any desired fault value. The smallest current transformer whose short-time rating will not exceed the maximum value of ground-fault current that may be anticipated is installed when utilizing the neutral ground strap type of system.

Figure 1-7 shows that the main ground-fault current path is from the transformer to the service, through the feeder to the fault, and back to the service over the equipment grounding conductor, then through the main bonding jumper where it returns to the transformer via the neutral conductor. The parallel circuit from the grounding electrode at the service to the grounding electrode is both a high-resistance and a high-reactance circuit. As a result, very little current will return through the earth. Some small

For additional information, visit qr.njatcdb.org Item #1046

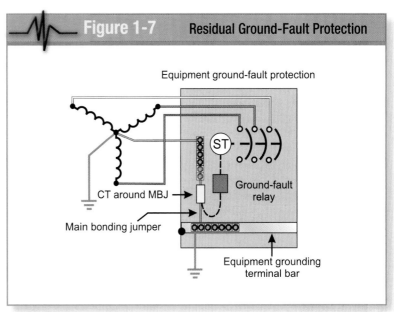

Figure 1-7. The residual ground-fault protection system is also known as the neutral ground-strap system.

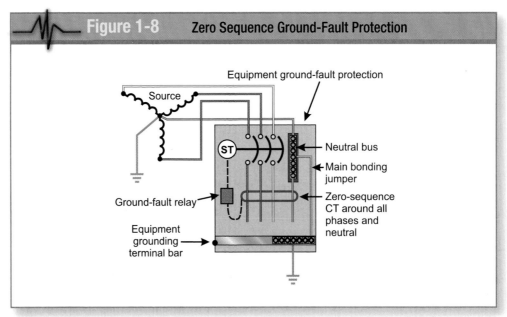

Figure 1-8. *The zero-sequence ground-fault protection system is the most popular.*

ground-fault current will be carried by the building's structural metal framing if it is in the circuit. It is preferable to adequately ground the building steel using the same grounding electrode as used for the service. This prevents the building's structural metal frame from rising to a dangerous potential above ground and serves as a grounding electrode as

required by 250.52(A)(2). Even though the building's structural metal framing represents a parallel path for fault current to flow, most of the current will flow back to the transformer through the neutral conductor because of the lower impedance of that path as compared to the impedance of the other parallel paths.

Zero-Sequence Transformer-Type System
The most popular kind of ground-fault protection system is the zero-sequence type. This system consists of a control relay, a shunt trip circuit breaker and a current transformer that is placed around all of the circuit conductors, including the grounded (neutral) conductor. **See Figure 1-8.**

The current transformer must be placed downstream from the main bonding jumper. The equipment grounding conductor does not pass through the current transformer.

A second version of a zero-sequence ground-fault protection system exists that utilizes an optional current transformer about the neutral conductor where one is used. **See Figure 1-9.**

The main difference between the two types of zero-sequence equipment is that the current transformers are generally built into the circuit breaker with an

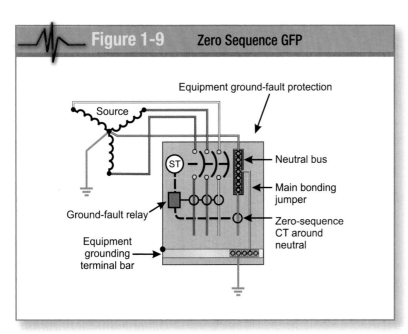

Figure 1-9. *The zero-sequence system can have an optional neutral sensing window.*

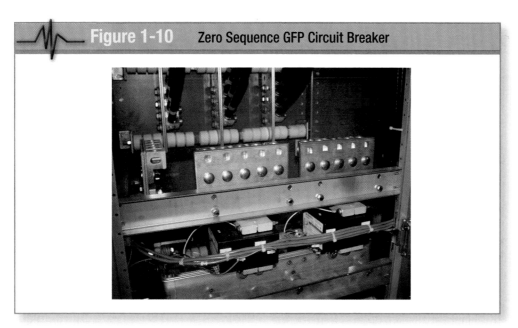

Figure 1-10. *The zero-sequence system can use a current transformer.*

external current transformer through which the neutral passes that is field installed. **See Figure 1-10.**

Often, these internal current transformers are used by the circuit breaker as a part of its operation system. This type of zero-sequence ground-fault protection system is sometimes referred to as a residual type. The system sums up the current through all four coils and considers any excess current residual.

Under normal operation, the vector summation of all phase and neutral (if used) currents approaches zero. This is due to the canceling effect of the currents in the conductors. A sensor (differential current transformer) around the phase and neutral conductors detects the current imbalance when a ground fault occurs. This is due to the fault current passing outside the current transformer, which sets up the imbalance which is detected by the system. The output of the sensor is proportional to the magnitude of the ground-fault current. This output is fed to a field adjustable ground-fault relay. Pickup ranges of 4 to 1,200 amperes are common. Time delay settings are available from $1\frac{1}{2}$ cycles to 36 cycles. When the ground-fault current exceeds a pre-selected level, the relay will activate the circuit-interrupting device, which

usually is a shunt trip circuit breaker, to open the circuit. **See Figure 1-11.**

Feeders

Where ground-fault protection is provided for operation of the service disconnecting means or feeder disconnecting means, an additional step of ground-fault protection must be provided in all next-level feeder disconnecting means

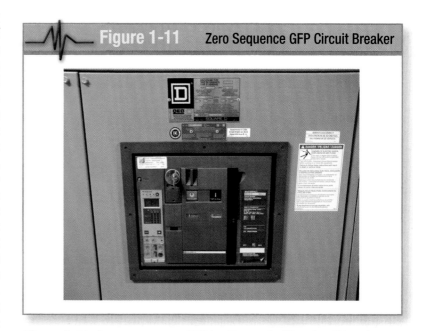

Figure 1-11. *A zero-sequence GFP is often part of the circuit breaker.*

downstream toward the load. Such protection must consist of overcurrent devices and current transformers or other equivalent protective equipment that will cause the feeder disconnecting means to open.

The additional levels of ground-fault protection must not be installed on the load side of an essential electrical system transfer switch. Nor should they be installed between the on-site generating unit(s) described in 517.35(B) and the essential electrical system transfer switch(es). Finally, these additional levels of protection must not be installed on systems that are not solidly grounded wye systems with greater than 150 volts to ground, but not exceeding 600 volts phase-to-phase.

Ground-fault protection of all equipment is generally required for solidly grounded wye electrical services of more than 150 volts to ground but not exceeding 600 volts phase-to-phase for each service disconnect rated 1,000 amperes or more. This protection is required for nominal 480Y/277-volt, 3-phase, 4-wire, wye connected systems. **See Figure 1-12**.

The maximum current setting for service disconnect ground-fault protection is 1,200 amperes, and the maximum time delay for operation is one second for ground-fault currents equal to or greater than 3,000 amperes.

Selectivity

Ground-fault protection for operation of the service and feeder disconnecting means must be fully selective. This means that the feeder device, but not the service device, shall open on ground faults on the load side of the feeder device. A separation between the service and feeder ground-fault tripping bands must be provided according to manufacturers' instructions. Operating time of the disconnecting devices shall be considered in selecting the time spread between these two bands to achieve 100% selectivity.

Coordinating the trip sequence of overcurrent and ground-fault protection devices in power distributions systems is necessary where selective coordination is required. Selective coordination can be accomplished by various combinations of overcurrent and ground-fault protective devices such as circuit breakers, fuses, or GFCI devices. GFCI devices differ from circuit breakers and fuses. A *ground-fault circuit interrupter (GFCI)* is a device intended for the protection of personnel that functions to de-energize a circuit or portion thereof within an established period of time when a current to round exceeds the values established for a Class A device. The *Code* adds the definition of

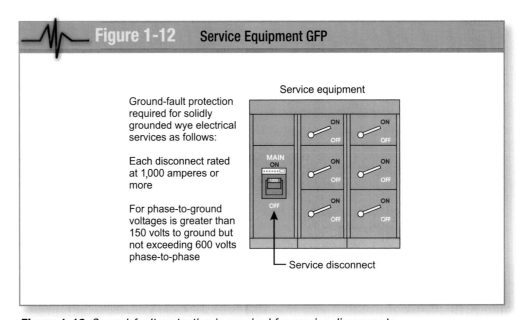

Figure 1-12   Service Equipment GFP

Ground-fault protection required for solidly grounded wye electrical services as follows:

Each disconnect rated at 1,000 amperes or more

For phase-to-ground voltages is greater than 150 volts to ground but not exceeding 600 volts phase-to-phase

Service equipment

MAIN ON

OFF

Service disconnect

Figure 1-12. Ground-fault protection is required for service disconnects.

a Class A device:

Informational Note: Class A ground-fault circuit interrupters trip when the current to ground is 6mA or higher and do not trip when the current to ground is less than 4 mA. For further information, see UL 943, *Standard for Ground-Fault Circuit Interrupters.*

Where ground-fault protective systems are installed, they should be designed so that they are selective to the point where the offending ground-fault event opens the closest upstream device from the fault event. Where overcurrent and ground-fault protection devices are selectively coordinated, they provide the benefits of restricting outages to the circuit or equipment closest to the ground-fault or short circuit event, by operating the local overcurrent or ground-fault protection device rather than causing the entire system to suffer a failure. **See Figure 1-13.**

The requirements for selective coordination are defined in Article 700.27 of the *Code*. They apply to all emergency systems. Article 517 includes specific rules for essential systems in health care facilities. It is important to remember that the emergency system in a hospital includes the life safety branch and the critical branch of the essential electrical system. Both of these branches are required to be selectively coordinated to minimize power failures only to those local areas directly affected by the faulted condition. Section 517.17 requires that all feeders having ground-fault protection be equipped with a second level of ground-fault protection installed in the subfeeder disconnecting means. It also requires this additional protection be selectively coordinated in order to provide the continuity of electrical service important to health care facilities where critical care and life support are essential for patients.

System coordination with circuit-breaker and fused distribution systems is readily accomplished by use of ground-fault protective devices. These ground-fault protective devices may be cascaded where economics of the design warrant doing so. The time-delay

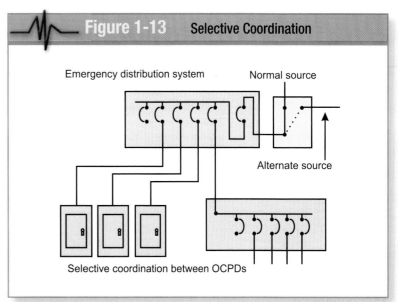

Figure 1-13. *Selective coordination is required for overcurrent devices used in emergency systems in accordance with 700.27.*

settings may become lower and lower as the device gets further downstream from the service, so that the last ground-fault protective device in the system may even be set for instantaneous trip. Similar coordination also can be obtained by locking out the relay or relays upstream from the sensor that experiences the fault first. **See Figure 1-14.**

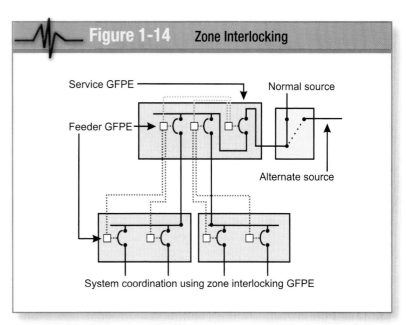

Figure 1-14. *System coordination can be accomplished using interlocking ground-fault protective devices at multiple levels.*

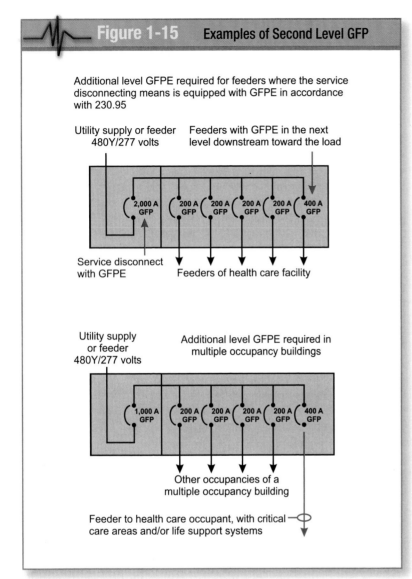

Figure 1-15   Examples of Second Level GFP

Additional level GFPE required for feeders where the service disconnecting means is equipped with GFPE in accordance with 230.95

Utility supply or feeder 480Y/277 volts

Feeders with GFPE in the next level downstream toward the load

2,000 A GFP   200 A GFP   200 A GFP   200 A GFP   200 A GFP   400 A GFP

Service disconnect with GFPE

Feeders of health care facility

Utility supply or feeder 480Y/277 volts

Additional level GFPE required in multiple occupancy buildings

1,000 A GFP   200 A GFP   200 A GFP   200 A GFP   200 A GFP   400 A GFP

Other occupancies of a multiple occupancy building

Feeder to health care occupant, with critical care areas and/or life support systems

Figure 1-15. A second level of ground-fault protection is required by 517.17(A) and (B).

Since it cannot be known where a ground fault may originate in the system, and since it is desirable to coordinate overcurrent protection and isolate the fault nearest to its point of origin, it is necessary to delay the action of tripping the service and upstream feeder ground-fault protection devices. A magnetic current relay is used to initiate a time delay relay to provide proper coordination and continuity of service consistent with safety.

Not all ground faults are high impedance. Should a low impedance fault occur, the normal main overcurrent device should be set to a time and current value that will allow the device to isolate the fault. Even though the ground-fault protective relay should be set at as low a value as possible, enough time delay should be set to permit the overcurrent device nearest the fault to open. The ground-fault protective device is an adjunct to, but does not replace, the overcurrent protective device. Ground-fault protective devices provide protection well below the normal ratings of the overcurrent device. **See Figure 1-15.**

The primary goal is to ensure that a ground fault event in the electrical system does not disconnect the service, but opens the device closest to the fault, thus isolating the offending circuit, while maintaining continuity of power to the rest of the facility.

This feeder protection is required to be 100% selective so that where a ground fault occurs downstream from the feeder overcurrent device, only the closest overcurrent device will open and the upstream devices will remain closed. This requirement is in place to prevent the blackout of a facility caused when a main could open for a fault that should be isolated to a single feeder. To achieve this coordination, manufacturers' recommendations of separation between the service and feeder tripping bands must be followed as per 517.17(C). **See Figure 1-16.**

Figure 1-17 illustrates a situation where the electrical system is delivered at more than 600 volts and has the service disconnecting means at that voltage level. A transformer then reduces the voltage of the feeder to the level where equipment ground-fault protection is required. The risk of destructive arcing burn downs is the same regardless of whether the system is a service or feeder. **See Figure 1-17.**

Fact

GFP selectivity changed for the 2011 *NEC*. No longer is there a 6-cycle selectivity requirement between tripping bands of GFP devices. Now the required selectivity must follow the manufacturers' recommendations to achieve 100% selectivity.

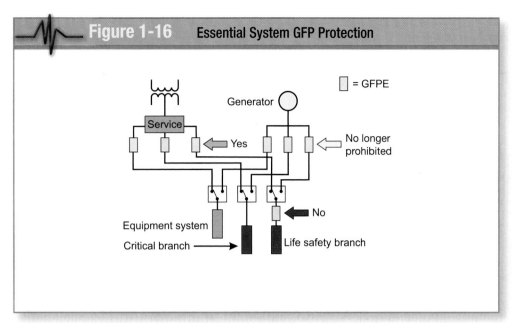

Figure 1-16. *Additional levels of ground-fault protection is NOT to be installed on the load side of an essential electrical system transfer switch(s).*

Care must be exercised to be certain that each new system created by a separately derived system has the protection required; and if it qualifies for equipment ground-fault protection by the voltage and ampacity level of the feeder and equipment, protection must be provided.

Testing

When equipment ground-fault protection is first installed, each level must be performance tested to ensure compliance with all the requirements for selectivity. Experience has shown that the majority of these systems do not operate properly, or at all, when first installed. All testing must be performed in full compliance with the manufacturer's written instructions. The instructions are usually furnished with the equipment. A vital part of the test is to remove the neutral disconnect link in the service equipment and test the neutral with a continuity tester or megohm meter to be certain that it is completely isolated from any grounding connections on the line side of the service disconnect. Accidental or intentional grounding connection to the grounded (neutral) conductor past the ground-fault protection device can render ground-fault protection systems ineffective. Written

records of the performance test are required to be made available to the authority having jurisdiction. Electrical system and equipment safety requires that the ground-fault protection equipment be properly installed and performance tested.

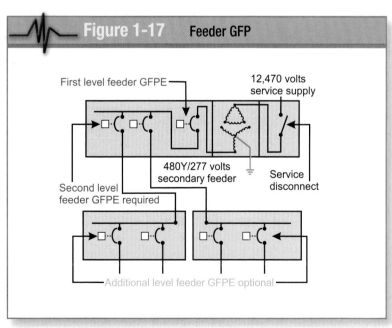

Figure 1-17. *Ground-fault protection is required for feeders.*

Summary

The wiring requirements that apply to patient care areas in all health care facilities in Part II of Article 517 significantly modify the general wiring rules in Chapters 1 through 4 of the *Code*, usually in a more restrictive fashion. Requiring redundant equipment grounding paths for branch circuits supplying patient care areas provides an added level of shock protection to a patient. Where ground-fault protection is installed in a service or a feeder that provides normal, life support, and critical care power, an additional level of ground-fault protection must be provided in the next level feeder downstream toward the load. This additional level of ground-fault protection for equipment, along with the requirements for selective coordination of overcurrent protective devices in the emergency system, provides greater reliability and continuity of electrical service in health care facilities.

Review Questions

1. Nursing homes and limited care facilities that provide only limited health care functions and primarily provide sleeping rooms for patients or residents are required to meet the wiring and protection requirements in Part II of Article 517.

   a. True

   b. False

2. A(n) __?__ supplies two or more receptacles or outlets for lighting and appliances and so forth in a health care facility.

   a. Individual branch circuit

   b. Multiwire branch circuit

   c. General purpose branch circuit

   d. Appliance branch circuit

3. Which of the following does not describe the purpose or characteristics of an equipment grounding conductor installed in any electrical system?

   a. It establishes a path to ground from conductive objects to essentially hold the conductive equipment or materials at or as close to the potential of the earth as possible.

   b. It provides an effective ground-fault current path to facilitate overcurrent device operation in the event of a ground fault in any of the ungrounded circuit conductors the circuit.

   c. It is generally required to be identified by the color green or green with one or more yellow stripes.

   d. It is used to connect the grounded conductor of a system and an electrical enclosure to the grounding electrode.

4. Which of the following wiring methods is not suitable for use in branch circuits serving patient care locations?

   a. Rigid metal conduit

   b. Rigid PVC Conduit

   c. Electrical metallic tubing

   d. Intermediate metal conduit

5. Wire-type equipment grounding conductors used in branch circuits serving patient care locations must meet which of the following requirements?

   a. They shall be insulated and shall be copper.

   b. They are permitted to be bare copper conductors.

   c. They are permitted to be insulated aluminum conductors.

   d. They are permitted to be eliminated as long as the conduit qualifies as an equipment grounding conductor in accordance with 250.118.

6. Where the branch circuit serving a patient care area is protected by an overcurrent device rated at 30 amperes, what is the minimum size insulated equipment copper grounding conductor required to be installed?

   a. 12 AWG

   b. 10 AWG

   c. 8 AWG Aluminum

   d. 14 AWG

7. **Which of the following health care locations prohibits the installation an isolated grounding-type receptacle connected to an isolated, insulated equipment grounding conductor?**

    a. Business office areas

    b. Computer rooms

    c. Nurse station area

    d. Patient care area

8. **The equipment grounding terminal bars of the normal and essential branch circuit panelboards serving the same individual patient vicinity are required to be bonded together with an insulated copper conductor not smaller than __?__ .**

    a. 12 AWG

    b. 10 AWG

    c. that required by 250.122, based on the rating of the overcurrent device protecting the feeder of the panelboards

    d. 8 AWG

9. **Where a 1,200-ampere, 480Y/277-volt service disconnect in a health care facility is equipped with ground fault protection, and the facility provides critical care and life support equipment, which of the following is true?**

    a. Equipotential bonding is required in the operations rooms.

    b. Fuses are required for the feeders.

    c. Breakers are permitted for the feeders.

    d. An additional level of ground-fault protection is required in the next level of feeder downstream toward the load.

10. **Where additional levels of ground fault protection are provided in feeders in accordance with 517.17(B), ground fault protection is not permitted in which of the following locations?**

    a. On the load side of an essential electrical system transfer switch

    b. Between the on-site generating unit(s) described in 517.35(B) and the essential electrical system transfer switch(es)

    c. On electrical systems that are not solidly grounded wye systems with greater than 150 volts to ground but not exceeding 600 volts phase-to-phase

    d. Ground fault protection for equipment is not permitted in any of the locations or conditions provided in choices (a) through (c).

11. **Where a service disconnect in a health care facility that provides critical care and life support is provided with ground-fault protection, which of the following sizes of feeders in the next level feeder downstream toward the load is required to have ground-fault protection?**

    a. 1,000 ampere

    b. 100 ampere

    c. 400 ampere

    d. All of the above

12. **Where ground-fault protection for equipment is included in the design of a hospital, when and where is it required to be performance tested?**

    a. In the electrical contractor's shop before being delivered to the job site

    b. In the manufacturer's facility before being delivered to the site

    c. When it is first installed on the job site

    d. Performance testing is not required for ground-fault protection equipment

13. **Which of the following areas in a health care facility are not classified as patient care areas?**

    a. Business offices or corridors

    b. Lounges and day rooms

    c. Dining rooms, or similar areas

    d. All of the above

14. **What is the minimum height above finished floor (AFF) that patient care area luminaires are exempt from the requirements of redundant grounding?**

    a. 9 1/2 feet

    b. 8 1/2 feet

    c. 7 1/2 feet

    d. 6 1/2 feet

Patient Care Wiring

Continuing with Part II of Article 517 of the *Code* and the importance of patient care area wiring, the focus is moved to the different types of patient care areas with their individual circuit(s), receptacle(s), and shock protection techniques within those wiring systems. Again, these requirements apply to all areas of patient care. The differences of grounding conductors and additional panelboard bonding are stressed in circuits to patient bed locations within general and critical care areas. For each type of patient care wiring, supply from both normal power system and the essential electrical system is required as well. The shock protection technique of isolated power systems will be introduced with more information provided in later chapters, and for wet procedure locations, high hazard and alternate shock techniques are covered. This chapter essentially completes our work within Part II, Wiring and Protection, of Article 517.

Objectives

» Determine the minimum number of receptacles and branch circuits for a patient bed location (in general care or critical care areas)

» Explain the need for hospital grade receptacles and for tamper-resistant receptacles (in pediatric locations)

» Describe how a patient care receptacle is identified in the field

» Explain the two methods used to protect patients from an electric shock in a wet procedure location

» Explain the benefit of providing an optional patient equipment grounding point within a critical care area

Chapter 2

Table of Contents

INTRODUCTION

Wiring methods, grounding practices, and general wiring practices used to install wiring in general care areas, critical care areas, and wet procedure location of health care facilities are particularly important in a health care facility. Remembering that Section 90.3 points out that *NEC* Chapter 5 (specifically Article 517) amends or modifies the requirements of Chapters 1 through 4 of the *NEC* is a critical concept to understanding health care installations. Part II of Article 517 focuses on patient safety by insisting upon enhanced reliability of circuits in patient care areas that may supply life support appliances. Also mandated by Part II of Article 517 is the reliability of the equipment grounding conductor, which provides automatic circuit disconnection should a fault occur. A redundant means of equipment grounding is required to ensure a low impedance ground path is available should one be lost. Also enhanced by Part II of Article 517 are the required quality and quantity of dedicated general care and critical care circuits and receptacles to serve the patient, and the reliability of those circuits even during power outages and other disasters.

Fact

The continuity of the equipment grounding conductor in patient care area wiring is extremely important and can not be broken, even if a receptacle is removed for maintenance.

GENERAL CARE AREAS

Health care facilities have areas devoted to different levels of patient care. There are the general areas, such as a patient's room. There are also critical care areas, where patients are monitored more closely and demands for special life-support equipment may be greater. Lastly, there are wet procedure locations (such as operating rooms) which have their own special requirements for both reliable electrical supply and protection from shock.

Patient Bed Location Circuit Requirements

In a general care location, each patient bed must be supplied by at least two branch circuits, one from the emergency system and one from the normal system. A 2011 change in the *NEC* [517.18(A)] now prohibits the use of multiwire branch circuits from serving patient bed locations in general care areas. This prohibition includes multiwire branch circuits supplied by the emergency system and supplied by the normal system. All branch circuits from the normal system must originate in the same panelboard. **See Figure 2-1.**

Three exceptions to this rule are as follows:

1. Branch circuits serving only special-purpose outlets or receptacles, such as portable X-ray outlets, are not required to be served from the same distribution panel or panels.
2. Patient bed locations in clinics, medical and dental offices, and outpatient facilities; psychiatric, substance abuse, and rehabilitation hospitals; sleeping rooms of nursing homes; and limited care facilities that are used for sleeping only and not patient care need not meet these requirements as long as they are wired in accordance with the other rules of Chapters 1 through 4 of the *Code*.

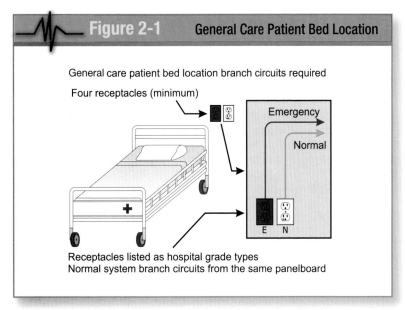

Figure 2-1 — **General Care Patient Bed Location**

General care patient bed location branch circuits required

Four receptacles (minimum)

Emergency

Normal

E   N

Receptacles listed as hospital grade types
Normal system branch circuits from the same panelboard

Figure 2-1. A minimum of two branch circuits are required for a general patient care bed location.

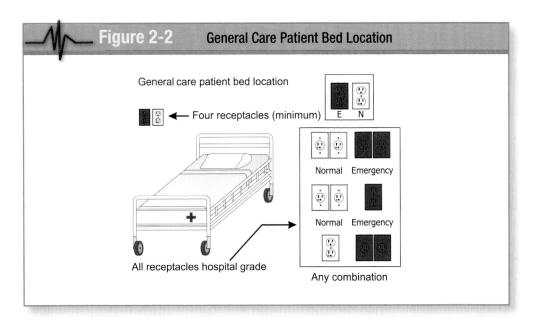

Figure 2-2. *A minimum of four receptacles are required for a general care patient bed location.*

3. A general care patient bed location served from two separate transfer switches on the emergency system is not required to have circuits from the normal system.

Patient Bed Location Receptacle Requirements

Each patient bed in a general care area must be provided with a minimum of four receptacles. These receptacles may be of the single, duplex, or quadplex type, or any combination of all three. All receptacles, whether four or more, must be listed "hospital grade" and be so identified. **See Figure 2-2.**

The grounding terminal of each receptacle must be connected to an insulated copper equipment grounding conductor sized in accordance with Table 250.122 of the *Code*.

There are two exceptions:

1. Psychiatric, substance abuse, and rehabilitation hospital rooms used only for sleeping, not patient care, need not meet these requirements as long as they are wired in accordance with the other rules of Chapters 1 through 4 of the *Code*.
2. Psychiatric security rooms are not required to have receptacles in them.

In existing facilities, a total, immediate replacement of existing non-hospital grade receptacles is not required. It is intended, however, that non-hospital grade receptacles be replaced with hospital grade ones upon modification of use, renovation, or as existing receptacles need replacement.

Hospital Grade Receptacles

The four receptacles required for a general care patient bed location must be listed "hospital grade." **See Figure 2-3.**

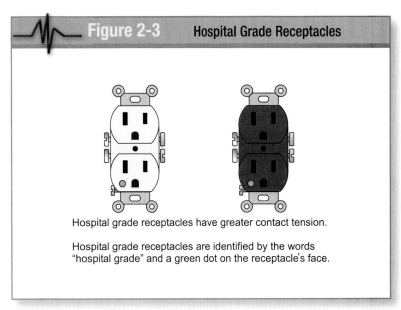

Hospital grade receptacles have greater contact tension.

Hospital grade receptacles are identified by the words "hospital grade" and a green dot on the receptacle's face.

Figure 2-3. *Hospital grade receptacles are required for general care patient bed locations.*

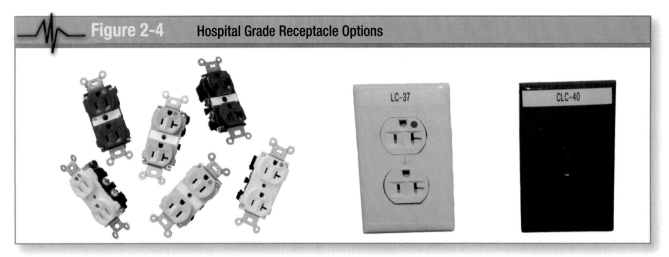

Figure 2-4   Hospital Grade Receptacle Options

Figure 2-4. Hospital grade receptacles are identifiable by their green dot.

For additional information, visit qr.njatcdb.org Item #1048

The primary difference between standard receptacles and hospital grade types is the tension between the blades of the attachment plug and the receiving contacts of the receptacles. Hospital grade receptacles are identified by the marking "hospital grade" and a green dot on the face of the receptacle. **See Figure 2-4.** Hospital grade receptacles required for patient bed locations should not be confused with listed "hospital use" receptacles that are for installations in hazardous anesthetizing locations as required by 517.61(B) and (C). See the UL General Information for Electrical Equipment

Directory category RTRT for additional technical information and marking requirements for receptacles that are listed as hospital grade types.

Receptacles that are not listed as hospital-grade must be tested at intervals not exceeding 12 months. Record keeping requirements are found in 8.5.3.1.3 of *NFPA 99*. A record must be maintained of the tests required by the chapter and of associated repairs or modification. At a minimum, this record must contain the date, the rooms tested, and an indication of which items have met or have failed to meet the minimum performance requirements. The *Code* is silent on the requirements for periodic performance testing of receptacles. *NFPA 99* addresses this requirement for receptacle testing in 4.3.4.1.3 in existing facilities only. This requirement applies to the receptacles installed at general care patient bed locations and critical care patient bed locations.

Pediatric Locations

Receptacles located within the rooms, bathrooms, playrooms, activity rooms, and patient care areas of pediatric wards must be listed tamper-resistant or use a listed tamper-resistant cover. **See Figure 2-5.**

Tamper-resistant receptacles are available in standard types, hospital grade, and as ground-fault circuit interrupter type receptacles. **See Figure 2-6.**

Tamper-resistant receptacles are constructed to require simultaneous insertion

Figure 2-5   Tamper-Resistant Hospital Grade Receptacle

Tamper-resistant receptacles provide protection against inserting foreign objects into the energized contacts.

Receptacles are identified by the words "Tamper-Resistant" or the letters "TR" on the receptacle so they are visible when the faceplate is removed.

These are available in standard types, hospital grade, and GFCI receptacle device types.

Figure 2-5. Listed tamper-resistant receptacles are required in pediatric areas.

of both the ungrounded and grounded conductor blades of an attachment plug to a minimum depth before the contacts of these receptacles will energize. This feature provides the safety if tampering occurs. Tamper-resistant receptacles are required to be identified by the words "Tamper-Resistant" or the letters "TR" where they will be visible after installation with the cover plate.

The UL Directory of Information for General Equipment (UL White Book) category KEVW provides additional product information about tamper-resistant receptacles in category RTRT.

Electrical Systems Less Than 120 Volts

Communications, signaling systems, data systems, and fire alarm systems are present in all health care facilities. **See Figure 2-7.**

Figure 2-6 **Tamper-Resistant Receptacle Options**

For additional information, visit qr.njatcdb.org Item #1049

Figure 2-6. Tamper-resistant receptacles come in standard types, hospital grade, and as ground-fault circuit interrupter types.

Figure 2-7 **Examples of Limited Energy Systems**

Figure 2-7. Limited energy systems are common in health care facilities.

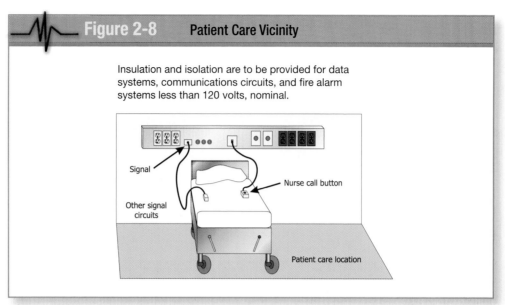

Figure 2-8   Patient Care Vicinity

Insulation and isolation are to be provided for data systems, communications circuits, and fire alarm systems less than 120 volts, nominal.

Signal

Other signal circuits

Nurse call button

Patient care location

Figure 2-8. *The patient care vicinity includes insulated signaling devices, such as nurse call system controls and other electronic controls that come in contact with the patient.*

Even limited energy circuits and systems can present shock risks to patients. However, where limited energy equipment and their associated circuits are designed, tested and listed for use in patient care areas of health care facilities, any risk of shock has been reduced substantially or eliminated.

Where low voltage and limited energy systems are installed in patient care areas, specific requirements in Part VI of Article 517 are applicable. Essentially, the goal is to isolate conductive surfaces in the patient vicinity. This is usually accomplished by installing insulated nurse call systems, television controls, signaling systems, and so forth within the patient vicinity. Also, to prevent misinterpretations concerning a statement of "… equivalent insulation or isolation…", the 2011 *NEC* has clarified that Class 2 and Class 3 signaling and communication systems and power limited fire alarm systems do not require additional grounding or physical protection unless specifically required by *NEC* Chapters 7 or 8. **See Figure 2-8.**

The Guide Information for Electrical Equipment (UL White Book) provides specific information about nurse call signaling systems and equipment under category NBRZ. This category is entitled "Hospital Signaling and Nurse Call Equipment." It covers units employed for general hospital signaling use, or to form part of a hospital nurse call signaling system. Equipment that has been found suitable for use in oxygen-enriched atmospheres or by patients undergoing oxygen therapy is identified as such in the listings. Only such listed equipment should be used in oxygen-enriched atmospheres or by patients undergoing oxygen therapy.

Permanently installed signal cable wiring from appliances in a patient location to remote apparatus or appliances is required to employ a signal transmission system that prevents hazardous grounding interconnection of the appliances. Common signal grounding wires (for example, the chassis ground for single-ended transmission) between appliances are permitted where all connected appliances are located within the patient vicinity and these appliances are served from and connected to the same reference grounding point. [517.82(A) and (B)]

Fact

Electrical equipment and other electrically conductive material likely to become energized must be installed in a manner that creates a low-impedance circuit, facilitating the operation of the overcurrent device. Also, it must be capable of safely carrying the maximum ground-fault current likely to be imposed on it from any point on the wiring system where a ground fault may occur to the electrical supply source.

CRITICAL CARE AREAS

Critical care areas are places where patients are more at risk and are monitored more closely than they would be in a general care room. These include such areas as intensive care units, coronary care units, angiography laboratories, cardiac catheterization laboratories, delivery rooms, and operating rooms. The requirements for specialized equipment are greater in critical care areas, and the effects of a power outage could be more devastating. **See Figure 2-9.**

Patient Bed Location Branch Circuits

Each patient bed location in a critical care area must be supplied by at least two branch circuits, one or more from the emergency system and one or more from the normal system. At least one branch circuit from the emergency system must supply an outlet(s) only at the bed

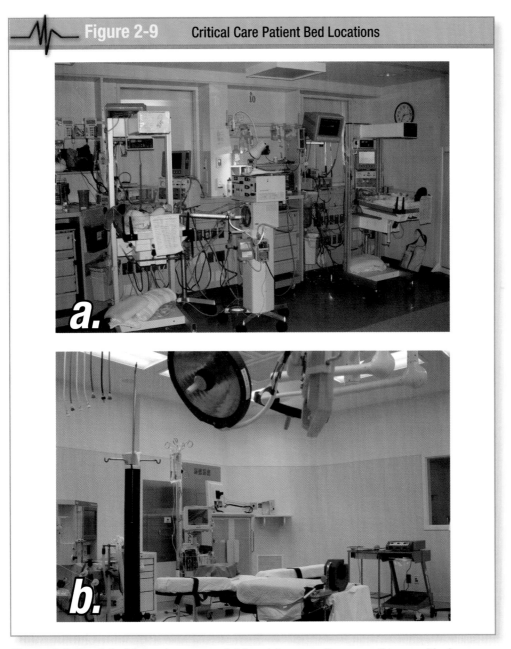

Figure 2-9   Critical Care Patient Bed Locations

Figure 2-9. The infant intensive care unit (a) and the operating room (b) are critical care patient bed locations.

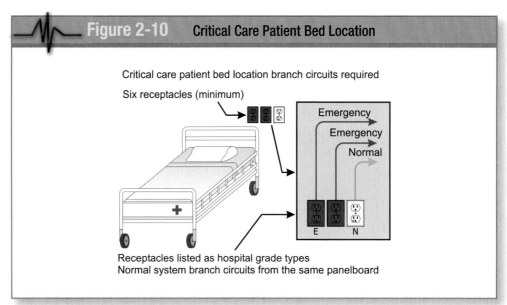

Figure 2-10. *A minimum of two branch circuits is required for a critical care patient bed location: one or more from the normal branch system and one or more from the critical branch system.*

location. All branch circuits from the normal system must be from a single panelboard. A 2011 change in the *NEC* [517.19(A)] now prohibits the use of multiwire branch circuits from serving patient bed locations in critical care areas. This prohibition includes multiwire branch circuits supplied by the emergency system and supplied by the normal system. Emergency system receptacles must be identified and also indicate the panelboard and circuit number supplying them. **See Figure 2-10.**

Two exceptions are as follows:

1. Branch circuits serving only special-purpose receptacles or

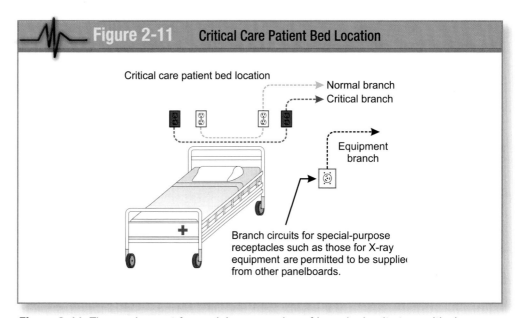

Figure 2-11. *The requirement for a minimum number of branch circuits to a critical care patient bed location does not apply to special purpose receptacles or outlets.*

equipment in critical care areas may be served by other panelboards. **See Figure 2-11.**

2. Critical care locations served from two separate transfer switches on the emergency system are not required to have circuits from the normal system.

Patient Bed Location Receptacles

Each patient bed location must be provided with a minimum of six receptacles, at least one of which must be connected to either of the following:

- The normal system branch circuit
- An emergency system branch circuit supplied by a different transfer switch from the other receptacles at the same location

These receptacles may be of the single, duplex or quadplex type, or a combination of both. All receptacles, whether six or more, must be listed "hospital grade" and be so identified. **See Figure 2-12.**

The grounding terminal of each receptacle must be connected to the reference grounding point by means of an insulated copper equipment grounding conductor.

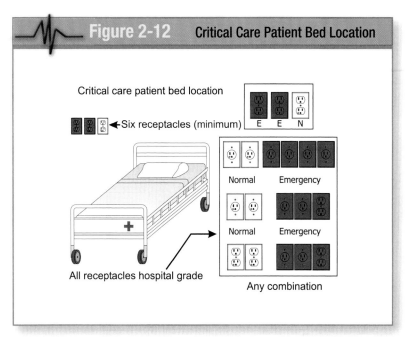

Figure 2-12. A minimum of six receptacles is required in a critical care patient bed location.

Emergency system receptacles installed at critical care patient bed locations are required to be specifically identified. The identification must indicate the panelboard and circuit supplying them. **See Figure 2-13.**

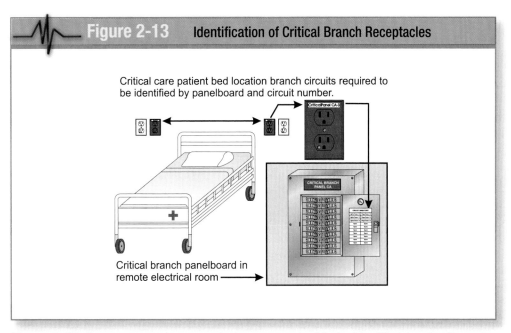

Figure 2-13. Emergency system receptacles are required to be identified by circuit number and panelboard.

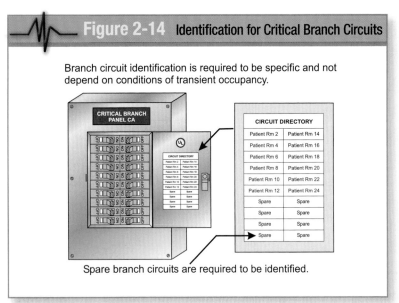

Figure 2-14 Identification for Critical Branch Circuits

Branch circuit identification is required to be specific and not depend on conditions of transient occupancy.

CRITICAL BRANCH PANEL CA

CIRCUIT DIRECTORY	
Patient Rm 2	Patient Rm 14
Patient Rm 4	Patient Rm 16
Patient Rm 6	Patient Rm 18
Patient Rm 8	Patient Rm 20
Patient Rm 10	Patient Rm 22
Patient Rm 12	Patient Rm 24
Spare	Spare
Spare	Spare
Spare	Spare
Spare	Spare

Spare branch circuits are required to be identified.

Figure 2-14. Panelboard circuit directories are required to be specific and not depend on conditions of transient occupancy. Spare overcurrent devices are required to be identified as spares.

Specific room identification is required at the panelboard location for each critical care branch circuit. **See Figure 2-14.**

Patient Care Vicinity Grounding and Bonding

Within the critical care area, an optional method of protecting patients from differences of potential and possible shock is the installation of a patient equipment grounding point. A *patient equipment grounding point* is a jack or terminal that serves as the collection point for redundant grounding of electric appliances serving a patient care vicinity or for grounding other items in order to eliminate electromagnetic interference problems. [99.3.3.141] **See Figure 2-15.**

The patient equipment grounding point, where supplied, may contain one or more listed grounding and bonding jacks. An equipment bonding jumper not smaller than 10 AWG must be used to connect the grounding terminal of all grounding-type receptacles to the patient equipment grounding point. The bonding conductor may be arranged centrically or looped as convenient. **See Figure 2-16.**

Where there is no patient equipment grounding point, it is important that the distance between the reference grounding point and the patient care vicinity be a short as possible to minimize any potential differences. Longer circuit runs increase the amount of capacitance in a circuit and can force potential differences to increase. If the option of a patient equipment grounding point is not employed in a patient care vicinity, it is a good practice to keep the branch circuit lengths as short as possible because the equipment grounding conductors of the branch-circuits are often the only direct path and connection to the reference grounding point in the supplying panelboard or isolated power system equipment. **See Figure 2-17.** An *isolated power system* is a system comprising an isolating transformer or its equivalent, a line isolation monitor, and its

Figure 2-15 Patient Equipment Grounding Point

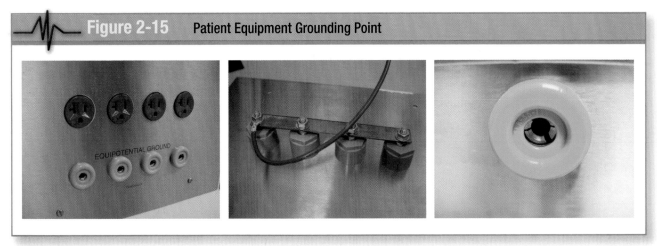

Figure 2-15. Listed grounding and bonding jacks are specially manufactured to serve as patient equipment grounding points.

ungrounded circuit conductors. Isolated power systems are discussed in more detail in Chapter 6 of this textbook, Isolated Power Systems.

Panelboard Grounding and Bonding

Where a grounded electrical distribution system is used and metal feeder raceway or Type MC or MI cable that qualifies as an equipment grounding conductor is installed, grounding of a panelboard or switchboard must be ensured by one of the following bonding means at each termination or junction point of the metal raceway or Type MC or MI cable:

- A grounding bushing and a continuous copper bonding jumper, sized in accordance with 250.122 of the *Code*, with the bonding jumper connected to the junction enclosure or the ground bus of the panel
- Connection of feeder raceways or Type MC or MI cable to threaded hubs or bosses on terminating enclosures
- Other approved devices such as bonding-type locknuts or bushings

This enhanced bonding requirement applies to all junction points (connection points at enclosures, junction and pull boxes, etc.) of the metal raceway or metallic MC or MI cable armor. Enhancing

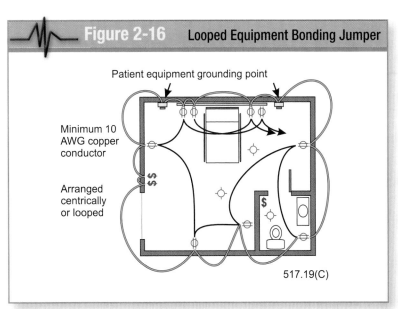

Figure 2-16. *A patient equipment grounding point and equipment bonding jumpers may be installed in a looped manner.*

the bonding requirements for feeders' equipment grounding means serves to ensure effective operation of branch circuit and feeder overcurrent devices in ground fault conditions. Since only one equipment grounding conductor is required for feeders in health care facilities to meet minimum *Code* requirements, the enhanced and more restrictive

Figure 2-17. *Equipment grounding conductors of the branch circuits are often the only direct path to the reference grounding point in the isolated power system.*

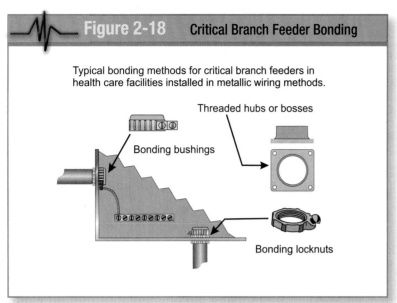

Figure 2-18   Critical Branch Feeder Bonding

Typical bonding methods for critical branch feeders in health care facilities installed in metallic wiring methods.

Threaded hubs or bosses

Bonding bushings

Bonding locknuts

Figure 2-18. Critical branch feeders require enhanced bonding and grounding.

bonding is essential. This bonding at termination points can be accomplished by a grounding (bonding) bushing and properly sized bonding jumper, threaded bosses or hubs, or other approved devices such as bonding-type locknuts or bushings. **See Figure 2-18.**

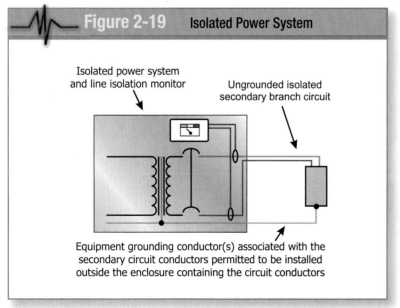

Figure 2-19   Isolated Power System

Isolated power system and line isolation monitor

Ungrounded isolated secondary branch circuit

Equipment grounding conductor(s) associated with the secondary circuit conductors permitted to be installed outside the enclosure containing the circuit conductors

Figure 2-19. Equipment grounding conductors installed with branch circuits supplied by isolated power systems are permitted to be installed outside the conduit or cable system for the branch circuit.

Additional Protective Techniques in Critical Care Areas

An optional protective technique is the use of an isolated power system. These are described in more detail under their own title later in this text. If used, the isolated power system equipment must be listed as isolated power equipment. The system must be installed in accordance with the requirements of 517.160, covered later in this text.

An isolated power system protects from shock by issuing audible and visual signals indicating leakage current. These signals must occur in a location where staff responsible for responding to them will notice them. So, for instance, in an operating room, the signals would generally have to occur within the room itself. In critical care areas where several patients are monitored from a central location, however, the line isolation monitor, which is the device that issues the signals, may be located in the nurses' station for the area being served.

Where an isolated ungrounded power source is used and limits the first-fault current to a low magnitude, the grounding conductor associated with the secondary circuit may be run outside of the enclosure of the power conductors in the same circuit. **See Figure 2-19.**

Although it is allowed to run the grounding conductor outside of the conduit, it is safer to run it with the power conductors. This provides better protection in case of a second ground fault.

Special-Purpose Receptacle Grounding

The equipment grounding conductor for special-purpose receptacles, such as mobile X-ray equipment, must be extended to the reference grounding points of branch circuits for all locations likely to be served from such receptacles. Where such a circuit is served from an isolated power system, the grounding conductor must not be required to be run with the power conductors. However, the equipment grounding terminal of the special-purpose receptacle must be connected to the reference grounding point.

WET PROCEDURE LOCATIONS

Any wet procedure location presents an added shock hazard. Bathrooms and sinks routinely have ground-fault circuit interrupters installed to mitigate the added risk. Therapeutic pools and tubs and their surrounding areas have special grounding and bonding requirements to reduce the risk of shock. Pools and tubs are covered under their own topic later in this text and in the *Code* under Part VI of Article 680.

The term *wet procedure location* refers to those spaces within patient care areas where a procedure is performed and that are normally subject to wet conditions while patients are present. These include standing fluids on the floor or drenching of the work area, either of which condition is intimate to the patient or staff. Routine housekeeping procedures and incidental spillage of liquids do not define a wet procedure location. There is an added complication with wet procedure locations that may include the operating room, where the floor may be covered in fluids, increasing the risk of shock to anyone standing in those fluids. Such areas cannot rely on ordinary overcurrent devices and GFCIs because these devices automatically shut off the power.

In fact, in patient rooms in critical care areas, if a sink or toilet is installed within the patient room, it does not require the use of a GFCI device because the risk of losing power to whatever life support equipment might be in the room outweighs the risk of someone being electrocuted at the sink.

In an operating room, such a loss of power could be fatal. Therefore, special provisions exist, such as isolated power systems..

Receptacles and Fixed Equipment

All receptacles and fixed equipment within the area of the wet procedure location must have ground-fault circuit-interrupter protection for personnel if interruption of power under fault conditions can be tolerated. If power interruption cannot be tolerated, then power must be supplied by an isolated power system.

Fact

Wet procedure locations in a health care facility include standing fluids on the floor or drenching of the work area, either of which condition is intimate to the patient or staff. However, routine housekeeping procedures and incidental spillage of liquids do not define a wet location.

There is an exception to this rule. Branch circuits supplying only listed, fixed, therapeutic, and diagnostic equipment may be supplied from a grounded service, single- or 3-phase system, provided both the following are true:

1. Wiring for grounded and isolated circuits does not occupy the same raceway.
2. All conductive surfaces of the equipment are connected to an equipment grounding conductor.

Isolated Power Systems

Where an isolated power system is utilized, the isolated power equipment must be listed as isolated power equipment and the system must be designed and installed in accordance with 517.160 (discussed later in this text).

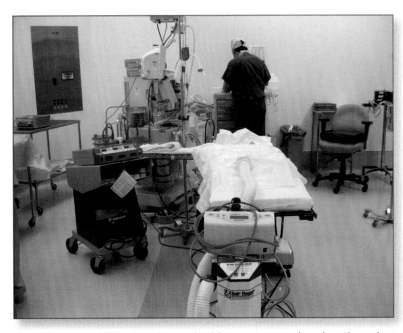

An isolated power system is required in a wet procedure location when power interruption cannot be tolerated.

Summary

The required wiring methods, grounding practices, and general installation practices used to install wiring in general care areas, critical care areas, and wet procedure locations are more restrictive than the general wiring rules in Chapters 1 through 4 of the *Code*. Even multiwire branch circuits are prohibited from being used to supply patient bed locations in general and critical care areas. Hospital grade receptacles are required at all patient bed locations, and tamper-resistant receptacles at all designated pediatric locations. Additional safety grounding is often provided in critical care locations such as operating rooms and wet procedure rooms. Also, enhanced grounding techniques are used in critical care feeders to ensure effective operation of the overcurrent devices.

Review Questions

1. Which of the following is responsible for determining the type of patient care and wet procedure locations in a hospital?
   a. The electrical contractor
   b. The patient
   c. The electrical engineer
   d. The governing body of the facility

2. Within the confines of a room, the perimeter of the patient vicinity extends __?__ in all horizontal directions around the perimeter of the patient bed.
   a. 4 feet
   b. 6 feet
   c. 8 feet
   d. 10 feet

3. Equivalent insulation and isolation to that required for the electrical distribution system is required to be provided in patient care areas for which of the following limited energy systems?
   a. Communications (telephone, intercom, etc.)
   b. Fire alarm systems
   c. Data systems and systems less than 120 volts
   d. All of the above

4. Which of the following locations does not meet the definition of a patient bed location?
   a. an inpatient bed in a hospital
   b. a procedure table in a doctor's examination office
   c. a procedure table in an operating room
   d. an inpatient bed in a critical care area (intensive care unit)

5. What is the minimum number of receptacles required at a patient bed location in a general care area?
   a. 2
   b. 3
   c. 4
   d. 6

6. What is the minimum number of receptacles required at a patient bed location in a critical care area?
   a. 2 single receptacles
   b. 3 duplex receptacles
   c. 4 single receptacles
   d. 2 duplex receptacles

7. What is the minimum number of branch circuits required to serve a patient bed location in a general care area?
   a. 1
   b. 2
   c. 3
   d. 4

8. The branch circuits supplying a patient bed location shall originate from which electrical system in a hospital?
   a. The normal system
   b. The emergency system
   c. One circuit from the normal system and one circuit from the emergency system
   d. The equipment system

9. **Listed hospital grade receptacles are identified by which of the following methods?**
   a. The color red
   b. The color blue
   c. An orange triangle on the receptacle face
   d. A green dot on the face of the receptacle and the words "hospital grade" on the mounting yoke of the receptacle

10. **A patient equipment grounding point serves which of the following purposes?**
   a. Serves as a collection point for redundant grounding of appliances
   b. Serves as a grounding means for other items to eliminate electromagnetic interference
   c. Both (a) and (b)
   d. Neither (a) nor (b)

11. **Where a patient equipment grounding point is installed in a critical care patient care location, the grounding terminals of all grounding-type receptacles shall be connected to the patient equipment grounding point with an equipment bonding jumper sized not smaller than   ? .**
   a. 12 AWG
   b. 10 AWG
   c. 8 AWG
   d. 14 AWG

12. **Which of the following does not qualify as a reference grounding point for a patient care location?**
   a. The equipment grounding bus in a service switchboard
   b. The ground bus of a panelboard that serves a patient care area
   c. The ground bus of an isolated power system panelboard
   d. Both (a) and (b)

13. **Receptacles in patient care areas are required to be tested in accordance with which of the following?**
   a. When first installed or after any repairs
   b. Where there is evidence of damage
   c. At intervals not exceeding 6 months
   d. Any of the conditions in (a) through (c)

14. **In wet procedure locations of health care facilities where interruption of power circuit(s) can be tolerated, an isolated power system is required to be installed.**
   a. True
   b. False

15. **Where special purpose receptacles, such as those for mobile X-ray equipment and other specialty medical appliances, are supplied by isolated power systems, the equipment grounding conductor for these branch circuits is also permitted to be installed on the outside of the enclosure (conduit or tubing system) in which the circuit conductors are installed. It is required to be connected to which of the following?**
   a. The equipment grounding terminal bar of the branch circuit panelboard supplying the circuit, or to the reference grounding point of an isolated ungrounded system
   b. Directly to a grounding electrode conductor
   c. Directly to a grounding electrode
   d. The grounded conductor terminal in a branch circuit panelboard

16. **Which of the following patient protective techniques used in critical care areas of hospitals are required, rather than optional?**
   a. Isolated power systems where interruption by a GFCI cannot be tolerated
   b. Patient vicinity grounding and bonding using a patient equipment grounding point
   c. Isolated power systems (in general)
   d. All of the above

17. **Which one of the following does not constitute a defined wet procedure location of a health care facility?**
   a. Areas that are normally subject to wet conditions while patients are present
   b. Areas that include standing fluids on the floor
   c. Areas that include drenching of the work area, either of which condition is intimate to the patient or staff
   d. Routine housekeeping procedures and incidental spillage of liquids

Essential Electrical Systems

An *essential electrical system* is a system comprised of alternate sources of power and all connected distribution systems and ancillary equipment, designed to ensure continuity of electrical power to designated areas and functions of a health care facility during disruption of normal power sources, and also designed to minimize disruption within the internal wiring system. [NFPA **99**:3.3.44]

Objectives

» Describe the basic operation of an essential electrical system during normal operation mode and during utility power failure mode

» Define essential electrical system requirements for hospitals

» Define essential electrical system requirements for nursing homes and limited care facilities

» Define essential electrical system requirements for other medical facilities, such as doctors' or dentists' offices

» Explain the requirements of essential electrical systems as they relate to branch circuits, feeders, and service, used in the three kinds of facility defined by the *Code*

Chapter 3

Table of Contents

For additional information, visit qr.njatcdb.org Item #1050

INTRODUCTION

Requirements for essential electrical systems are defined by the *NEC* in Part III of Article 517 and by Chapter 4 of *NFPA 99*. Sections 517.25 and 517.26 define the scope and application of the essential electrical system. After this definition of scope, Article 517, Part III is broken down into three basic categories based on occupancy:

1. Sections 517.30 through 517.35 apply to the essential electrical systems specifically required for hospitals.
2. Sections 517.40 through 517.44 apply to the essential electrical systems specifically required for nursing homes and limited care facilities.
3. Section 517.45 provides the general requirements for essential electrical systems in all other types of health care facilities.

Each of the three categories defined by the *Code* can be further divided into different subsystems, such as emergency systems and equipment systems. In addition, within every essential system, there will be specific and sometimes different requirements for branch-circuit and feeder wiring methods.

The essential electrical system is not simply a list of installed parts and pieces to generate electricity and deliver that electricity to a list of equipment. It is not just a list of mandatory code requirements either.

Rather, the reliability of an essential electrical system depends on many important selections, such as:

- The electrical equipment
- The physical placement of the equipment
- The type and quality of the wiring method
- The selected locations and the final circuit length
- The overall quality of the components
- Adherence to all *Code* requirements
- The ability of the components to work together as an efficient, safe, coordinated, and reliable system

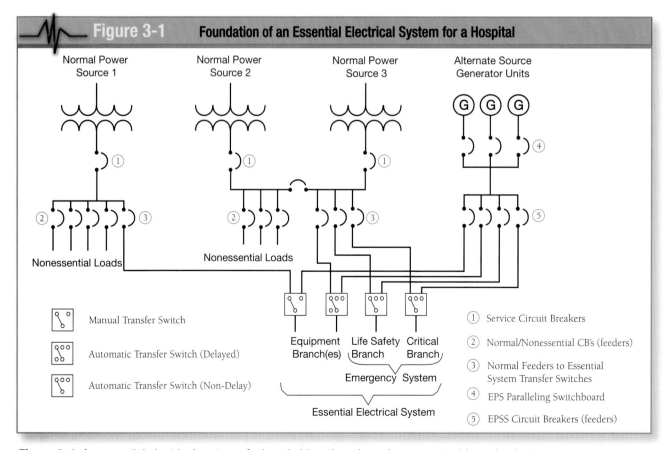

Figure 3-1. An essential electrical system of a hospital has three branches supported by redundant power sources.

Above all, an essential electrical system should function seamlessly whenever it is called upon to operate. The role of the electrical worker cannot be overstated in providing a safe and reliable essential electrical system.

HOSPITALS

The term *hospital* is defined as a building or portion thereof used on a 24-hour basis for the medical, psychiatric, obstetrical, or surgical care of four or more inpatients. [**101**:3.3.124] The term includes general hospitals, mental hospitals, tuberculosis hospitals, children's hospitals, and any such facilities providing inpatient care.

A vital element of electrical power systems designs in health care facilities involves designing for uninterrupted electrical power or continuity of service to the facility. For obvious reasons, hospitals tend to have the most urgent requirements for uninterrupted service. Any facility that performs medical procedures for which a loss of power would be catastrophic needs to be designed to avoid that catastrophe. Hospitals are the most demanding of such facilities and have the most stringent requirements. Extensive coverage of the code-related design criteria and the electrical requirements for essential electrical systems in hospitals may be found in *NFPA 99*, Section 4-4 as well as *NFPA 110, Standard for Emergency and Standby Power Systems*. **See Figure 3-1.**

Service

At the service level, electricity is normally provided to a hospital by the local electric utility. Figure 3-1 shows the normal (utility supplied) power sources and the alternate power source supplying the facility. An *alternate power source* is one or more generator sets, or battery systems where permitted, intended to provide power during the interruption of the normal electrical services or the public utility electrical service intended to provide power during interruption of service normally provided by the generating facilities on the premises.

As for any facility, the various utility supplied service(s) will be designed to

meet all the expected and any anticipated future power requirements. Because hospitals have certain essential functions that cannot be allowed to fail, the power supply must be designed and built to a high degree of reliability. Therefore, hospitals require a second source of supply capable of supporting their essential functions. This second source shown in Figure 3-1 may be a second utility line and, more likely today, a second utility source plus an onsite electrical generator(s).

In some rare cases where the local utility is not a consistently reliable source of electricity, however, the primary service may actually be the hospital's own generators, with a second set of generators and the local utility as the backup.

For additional information, visit qr.njatcdb.org Item #1051

517.35 Sources of Power:

(A) Two Independent Sources of Power. Essential electrical systems shall have a minimum of two independent sources of power: a normal source generally supplying the entire electrical system and one or more alternate sources for use when the normal source is interrupted. [**99**:4.4.1.1.4]

(B) Alternate Source of Power. The alternate source of power shall be one of the following:

(1) Generator(s) driven by some form of prime mover(s) and located on the premises

(2) Another generating unit(s) where the normal source consists of a generating unit(s) located on the premises

(3) An external utility service when the normal source consists of a generating unit(s) located on the premises

(4) A battery system located on the premises [**99**:4.4.1.2]

Generally, full operation and load transfer from the normal power source to the alternate power source (for most of the essential system) must occur within 10 seconds (see Section 517.31). While not specifically required by the *NEC*, UPS systems that virtually eliminate interruption or loss of power are most important to a hospital facility where computer-based medical records and images are handled. UPS systems are finding their way into a host of other system power supplies as well.

Feeder

From the initial point(s) of service, or source of power, the essential electrical system feeds into a collection of transfer

Figure 3-2   Foundation and Branches of an Essential Electrical System

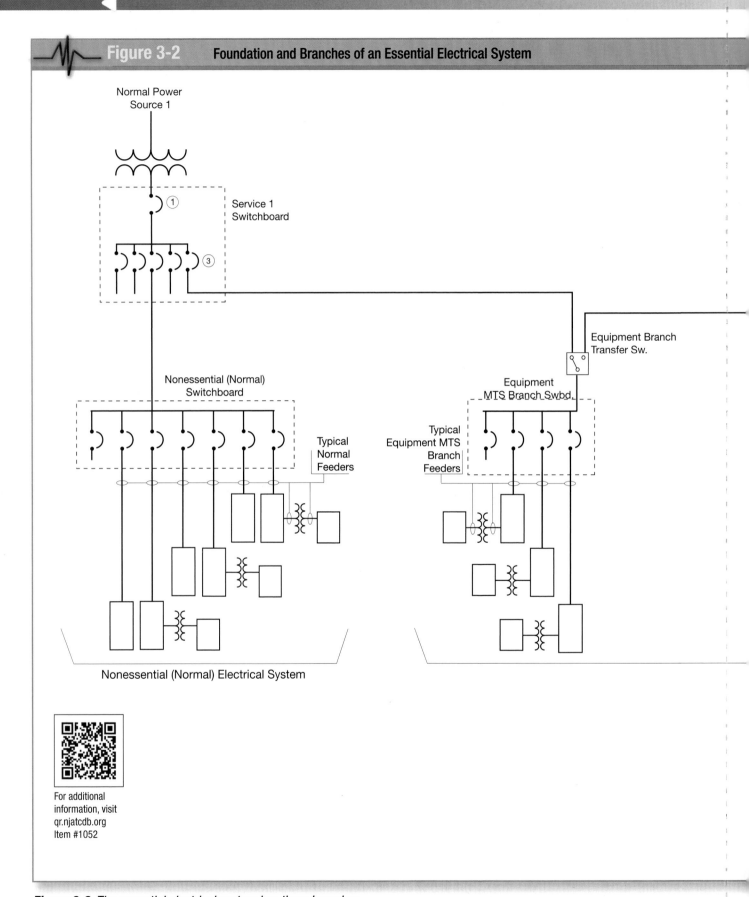

Figure 3-2. *The essential electrical system has three branches.*

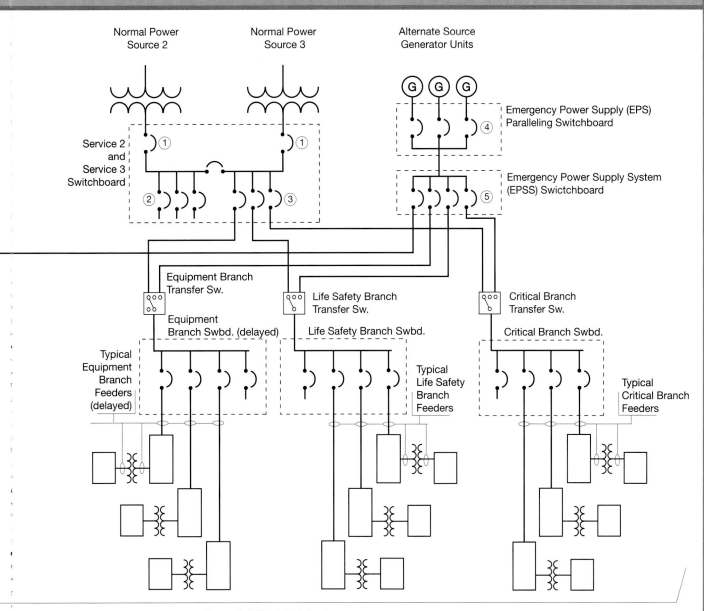

Essential Electrical System

① Service Circuit Breakers

② Normal/Nonessential CB's (feeders)

③ Normal Feeder to Essential System Transfer Switches

④ Emergency Power Supply (EPS) Paralleling Switchboard Feeder to Emergency Power Supply System (EPSS) Swictchboard

⑤ Emergency Power Supply System (EPSS) Feeder to Essential System Transfer Switches

 Manual Transfer Switch

 Automatic Transfer Switch (Delayed)

 Automatic Transfer Switch (Non-Delay)

switches whose purpose is to transfer the branches they serve between power sources as needed.

Figure 3-1 shows two feeders supplying each of the four transfer switches. Each transfer switch is supplied by a normal (utility supplied) feeder as well as a (generator) feeder supplied from the alternate source generator. Each automatic transfer switch is able to sense loss of power from a utility source, start the generator, wait for the proper voltage from the generator, and then immediately transfer load (or switch) from the utility source to the generator source, thereby restoring power to the downstream electrical equipment. This arrangement provides a level of power supply redundancy to essential electrical loads.

Feeders are sized based on the loads they serve, and size is calculated in accordance with the requirements for all feeders, regardless of whether they exist in hospitals or not. These requirements are covered in Articles 215 and 220 and, according to Section 90.3, may be amended in Article 517.

Branches

An essential electrical system is separated into three subsystems or branches. These branches are life safety, critical, and equipment. Each branch of the essential electrical system is named for the electrical load it serves, and each branch consists of both feeders and branch circuits. **See Figure 3-2.**

Life safety and critical systems are frequently grouped under the larger category of "Emergency System." Such grouping allows one to treat requirements that frequently apply to both systems together rather than having to say the same thing twice.

Equipment Branch

The equipment branch of a hospital essential electrical system is required to provide power for major electrical equipment that is necessary for patient care and basic hospital operation. Loads that would be included in the equipment branch often include, but are not limited to, ventilation systems (environmental air), suction systems serving medical and surgical functions, compressed air systems, and sump pumps.

The equipment system must be equipped with at least one transfer switch. Although not required, most hospitals include more than the minimum of one transfer switch in each branch to enhance reliability and to separate or handle large loads. Using multiple transfer switches can provide reasonable assurances that during power failure at locations downstream at the branch circuit level, the majority of the hospital remains operational on the normal system. As an absolute minimum *Code* requirement, 517.30(B) (4) requires at least one transfer switch in each branch. **See Figure 3-3.**

There is one exception. 517.30(B)(4) permits a single transfer switch to supply the emergency system and the equipment system in a small hospital with a maximum demand on the essential electrical system of 150 kVA or less. **See Figure 3-4.**

Certain electrical equipment must be connected to the equipment system, and that system must be able to transfer from normal to the alternate power source if

For additional information, visit qr.njatcdb.org Item #1053

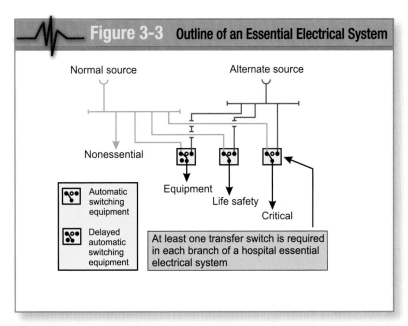

Figure 3-3. A minimum of one transfer switch is required for each system branch (NEC Informational Note Figure 517.30, No. 1).

the normal source is totally or partially interrupted or experiences drastic power quality issues. For some of these systems, the transfer must be automatic. For some, the transfer can be delayed and/or even manually switched to an alternate source. For example, in Figure 3-1, there are three different symbols for transfer switches. On the equipment branch, one is a delayed operation transfer switch and one is a manual transfer switch. If required, but not present in Figure 3-1, an automatic transfer switch could be used if certain equipment required immediate (or non-delayed) alternate power. Article 517.34(A) requires the following systems to delay transfer automatically, at appropriately defined time-lag intervals:

1. Central suction systems serving medical and surgical functions, including any associated controls
2. Sump pumps and equipment required to operate for the safety of major apparatus, including associated controls for such equipment
3. Compressed air systems including associated controls
4. Smoke control systems and stair pressurization systems
5. Kitchen hood supply or exhaust air systems, or both, that are required to be operational during a fire event in or under the hood
6. Supply, return, and exhaust ventilating systems for all of the following:
   - Airborne infectious/isolation rooms
   - Protective environment rooms
   - Laboratory room fume exhaust hoods
   - Nuclear medicine areas where radioactive materials are used
   - Ethylene oxide evacuation
   - Anesthesia evacuation
7. Supply, return, and exhaust ventilating systems for operating and delivery rooms

There is an order of priority established by the *Code* for the equipment branch. It is generally understood that the equipment system supplies loads that are related to continued operation of the facility, but not emergency branch loads. It is logical that the equipment branch must be designed to restore power after the life

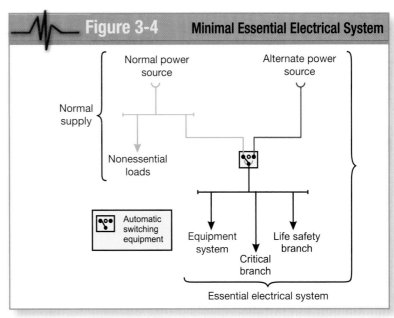

Figure 3-4. *A single transfer switch is permitted in small facilities under 150 kVA.*

safety and critical branch loads are restored and working.

Where equipment system loads are determined to be necessary without delayed automatic connection, they are permitted to be connected to the critical branch system. Some equipment systems may also be permitted to be connected to the life safety system. One example of such a system may be damper controls for a smoke evacuation system. Whenever this system is required to operate, this system must start up and work immediately. This permission is found in 517.32(C)(3).

Also, in the process of hospital construction or remodeling, it is normal for circuits to be reassigned to different branches. This can sometimes be confusing to the electrical workers because the actual operation of the many pieces of equipment is not always obvious during construction, but it is normal for hospital construction.

The *Code* also allows power to be restored in sequence, rather than all at once, in order to account for generator capacity concerns. In other words, if the generator could become overloaded by immediate connection of many equipment loads or a few large equipment loads, as determined through appropriate engineering studies, the *Code* permits a delayed automatic transfer switch to be used. However, instead of an immediate transfer and restoration of power, the transfer is delayed or slowed down as appropriate, thereby allowing the generator more time to stabilize. This function is permitted in 517.34(A) Exception. These time-delay issues of restoring electrical power to specific equipment are very important issues reserved for hospital administration and electrical engineers. They are not decisions for electrical workers.

517.34(B) lists the following equipment for either delayed automatic or manual connection to the alternate power source:

1. Heating equipment that provides heating for locations such as operating rooms, delivery rooms, labor and recovery rooms, intensive care units, coronary care units, nurseries, infection/isolation rooms, emergency treatment spaces, general patient rooms, fire pumps, and pressure maintenance pumps for water-based fire protection systems
2. Elevators necessary to provide service to patient, surgical, obstetrical, and ground floors during periods of interruption of normal power to the hospital.
3. Hyperbaric and Hypobaric facilities
4. Electrically operated doors
5. Minimal electrically heated autoclaving equipment
6. Controls for the equipment listed in Section 517.34
7. Other selected equipment as determined by the hospital governing body

As the architect, the engineer, and the hospital administration review electrical drawings for a new project, many branch circuits are added or moved among the normal system, the critical, and equipment branches of the essential electrical system. This may also happen long after construction is finished and the hospital determines that the facility does not function as expected during a power loss, emergency, or other power-related crisis.

Generally, the equipment branch will experience the most circuit changes during the first few years of operation.

Fact

Automatic transfer switches provide "…immediate restoration of service." The term *Immediate Restoration of Service* is defined in *NFPA 99*, 3.3.77 as "… Automatic restoration of operation with an interruption of not more than 10 seconds." Where a specific appliance or procedure cannot tolerate even this brief interruption of 10 seconds, a quicker acting Uninterruptable Power Supply (UPS) system can provide a better solution.

Emergency System

The *emergency system* in a hospital is defined as a system of circuits and equipment intended to supply alternate power to a limited number of prescribed functions vital to the protection of life and safety. [99:3.3.41] **See Figure 3-5.**

The emergency systems installed in buildings that are not health care facilities are typically installed in accordance with the requirements of Article 700 for occupant safety during power failures and allow for safe egress of occupants. These must be in operation for a limited amount of time, usually 1½ to 2 hours. Essential systems in hospitals, on the other hand, are much more comprehensive and necessary not only for safety reasons, but also for providing critical power systems continuity in the interest of the health and welfare of those patients that must remain in the care of the hospital. Patients who are confined to their beds, in addition to being unable to evacuate, might be attached to life-sustaining equipment, and their removal from the building might put them in grave danger. For these reasons, hospitals are designed to accommodate the defend-in-place strategy, whereby occupants are relocated to a safe location on the same floor rather than being evacuated. The safe locations are created by subdividing the floors of the building into two or more smoke compartments or fire compartments, separated by specially constructed walls designed to limit the transfer of smoke or restrict the spread of fire from one side to the other. These rooms must continue to receive power from the emergency system. Only life safety (essential) emergency system circuits and critical patient care circuits are permitted to be connected to the emergency system. [517.30(B)(2)] Also, where a hospital also serves a homeland security or civil defense function, the life safety system may be a more extensive system extending even into conference rooms and other designated areas of importance.

The emergency system of a hospital is viewed as a subsystem of the essential electrical system consisting of two important and separate branches: the life

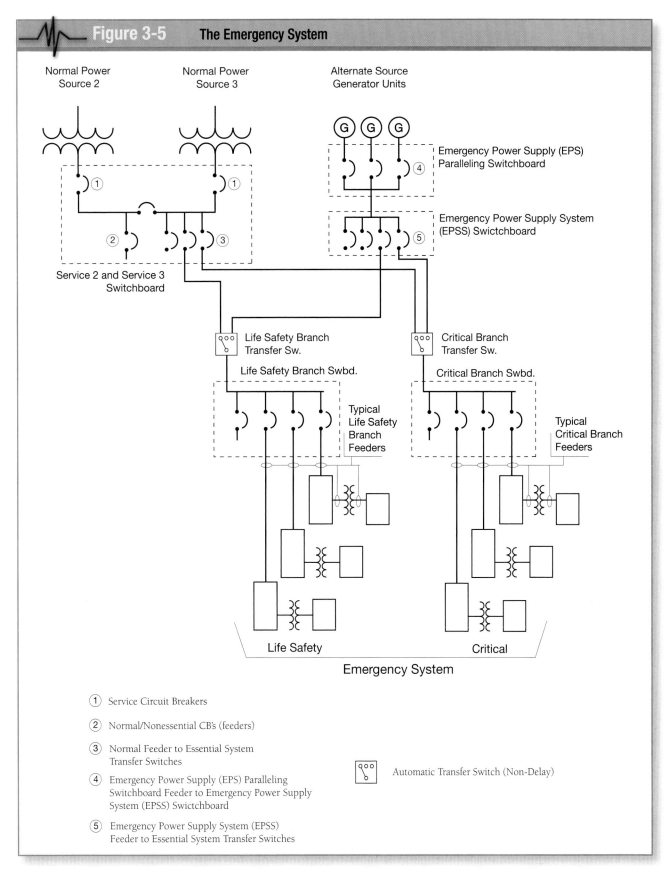

Figure 3-5   The Emergency System

Normal Power Source 2

Normal Power Source 3

Alternate Source Generator Units

Emergency Power Supply (EPS) Paralleling Switchboard

Emergency Power Supply System (EPSS) Swictchboard

Service 2 and Service 3 Switchboard

Life Safety Branch Transfer Sw.

Critical Branch Transfer Sw.

Life Safety Branch Swbd.

Critical Branch Swbd.

Typical Life Safety Branch Feeders

Typical Critical Branch Feeders

Life Safety

Critical

Emergency System

① Service Circuit Breakers

② Normal/Nonessential CB's (feeders)

③ Normal Feeder to Essential System Transfer Switches

④ Emergency Power Supply (EPS) Paralleling Switchboard Feeder to Emergency Power Supply System (EPSS) Swictchboard

⑤ Emergency Power Supply System (EPSS) Feeder to Essential System Transfer Switches

Automatic Transfer Switch (Non-Delay)

Figure 3-5. The emergency system consists of the life safety and critical branches of the essential electrical system.

Figure 3-6 **Automatic Transfer Switch (ATS)**

Figure 3-6. *Wiring separation is not required in transfer switch enclosures.*

are never allowed to share the same raceway, enclosure, box, or cabinet with normal or other non-emergency system wiring. The reason for these strong separation requirements is to prevent a fault or failure in any one system from causing a fault or failure in another system. Any failure in a normal or non-emergency system must not be allowed to interrupt the ability of the emergency system to function properly.

However, these branch separation requirements do require some common sense exceptions. In accordance with 517.30(C)(1), there are four very limited conditions in which the wiring of the life safety branch and the critical branch system are permitted in the same enclosure, box, raceway, or cabinet with each other:

1. Transfer equipment enclosures. **See Figure 3-6.**
2. Exit or emergency system luminaires supplied from two sources
3. Common junction boxes of emergency or exit luminaires supplied from two sources
4. Two or more emergency circuits supplied from the same system branch and same transfer switch.

safety branch and critical branch. Before dealing with each system individually, however, some common characteristics should be considered. The following four subjects are somewhat shared by both the life safety branch and the critical branch:

1. Circuit wiring separation requirement
2. Mechanical protection
3. Ten second transfer
4. Receptacle identification

Circuit Wiring Separation Requirements. 517.30(C)(1) requires that the life safety and critical branches of a hospital emergency system be kept entirely independent of all other wiring and systems and kept separate from each other as well. "Independent and separate" means that these two branches cannot share a common raceway, enclosure, box, or cabinet with each other. It also means that both systems

Mechanical Protection. Due to the importance of each branch, the *Code* mandates certain protection techniques to prevent physical damage. 517.30(C)(3) requires that the life safety and critical branches of a hospital emergency system be mechanically protected. These required protection techniques must be included at both the feeder circuit and the branch circuit levels including life safety and critical branch circuits that serve patient care areas. **See Figure 3-7.**

The most popular method of complying with the requirements of providing mechanical protection of the life safety and critical branches of the emergency system in a hospital is to install the conductors in a nonflexible metal raceway system. Most designers will specify the nonflexible metal raceway wiring method in their designs, meeting the intent of 517.30(C)(3)(1). The entire list of permitted wiring methods for the emergency

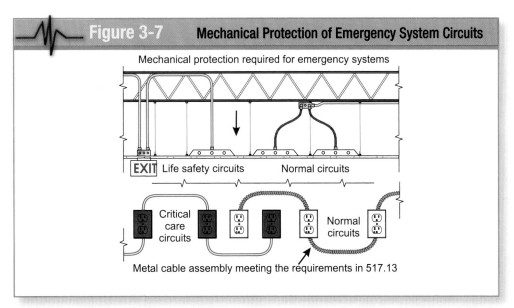

Figure 3-7. *Mechanical protection is required for the life safety and critical branches of a hospital emergency system.*

system is provided in 517.30(C)(3)(1) through (5) as follows:

1. Nonflexible metal raceways, Type MI cable, or Schedule 80 PVC conduit
2. Schedule 40 PVC conduit, flexible nonmetallic or jacketed raceways, jacketed metallic cable assemblies listed for installation in direct contact with the concrete, that are encased in not less than 50 mm (2 in.) of concrete. *Note: Nonmetallic wiring methods do not satisfy the redundant equipment grounding conductor rules in Section 517.13.*
3. Listed flexible metal raceways and listed metal sheathed cable assemblies in the following types of installations:
   a. Where used in listed prefabricated medical headwalls
   b. Where used in listed office furnishings
   c. Where fished into existing walls and ceilings that are not otherwise accessible and the fished wiring will not be subject to any physical damage
   d. Where necessary for flexible connection to equipment
      Note: Many interpret this provision as an allowance for flexible whips to luminaires in suspended ceilings.

Be careful with this Code rule; this section specifically indicates that flexible connection is necessary. A good way to look at this is to ask the question: Can this equipment be hard wired, or is there no other way than to wire it with flexible wiring methods? This approach will usually ensure compliance with this section.

4. Flexible power cords of utilization equipment or appliances connected to the emergency system
5. Cables for Class 2 or Class 3 systems permitted by Part VI of this Article, with or without raceways.

 Fact

Mechanical protection over and above that required in ordinary commercial occupancies is required by the *Code* to fundamentally reduce the risk of damage to the equipment grounding conductors within each branch circuit and to add reliability to the continued safe operation of the circuit. Physical stress on a cable can loosen the cable from its connector, thereby eliminating one of the equipment grounding conductor(s).

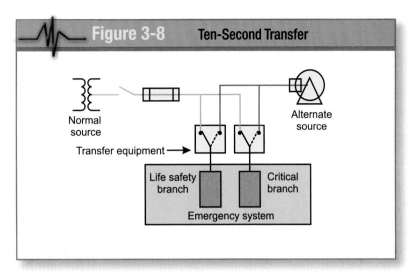

Figure 3-8. *The emergency system in a hospital is required to be connected to the alternate power source within 10 seconds of a power loss.*

Ten-Second Transfer. The *Code* prescribes that both the life safety branch and the critical branch of the emergency system in a hospital must be connected to the alternate power source within 10 seconds of a power loss condition. This includes all loads connected to the life safety and the critical branches. **See Figure 3-8.**

Receptacle Identification. 517.30(E) provides identification requirements for receptacles supplied by the emergency system of a hospital. The most common method of identification used today is the color red, although the *Code* does not specifically require red as the color to be used. The rule calls for the receptacle cover plate or the receptacles themselves to have a distinctive color or marking so as to readily identify them as connected to the emergency system. **See Figure 3-9.**

This requirement is important for operational staff especially when a facility suffers a loss of power. Identifying the receptacles as supplied from the emergency branch makes it easy to know which receptacles will still have power. Important life support equipment can be plugged into one of these receptacles, if it is not already.

Life Safety Branch

The *life safety branch* of a hospital (or any health care facility) is defined as a branch of the emergency system consisting of feeders and branch circuits, meeting the requirements of Article 700 and intended to provide adequate power needs to ensure safety to patients and personnel, and that are automatically connected to alternate power sources during interruption of the normal power source. [99:3.3.96] **See Figure 3-10.**

Other facilities, non-health care facilities, also maintain life safety systems, but those systems are in place to operate for a minimum amount of time to allow for vacating a building or structure in a safe manner. Hospitals have a similar need, but they also need to operate life-support equipment throughout the power outage, not simply long enough for people to vacate a building. The power demands of that equipment are greater than those of simple stairway exit lights and the demands are likely to continue for an extended period of time. As mentioned earlier, a hospital's life safety system may need to support a defend-in-place strategy, where patients who cannot be evacuated are moved to designated safe locations where extra smoke and fire protection is in place. Patients may be attached to life-sustaining equipment that must continue to function during the emergency. The life

Figure 3-9 — Receptacle Identification

Receptacles or cover plates are required to be identified using a distinctive color so as to be readily identifiable.

EMERG C-5 → **EMERG C-3**

Readily identifiable as a receptacle connected to the emergency system using red (color not mandated)

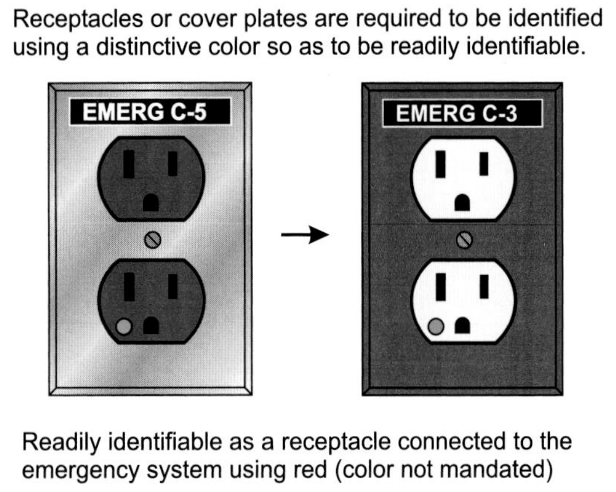

Figure 3-9. *Receptacles or cover plates are required to be identified using a distinctive color.*

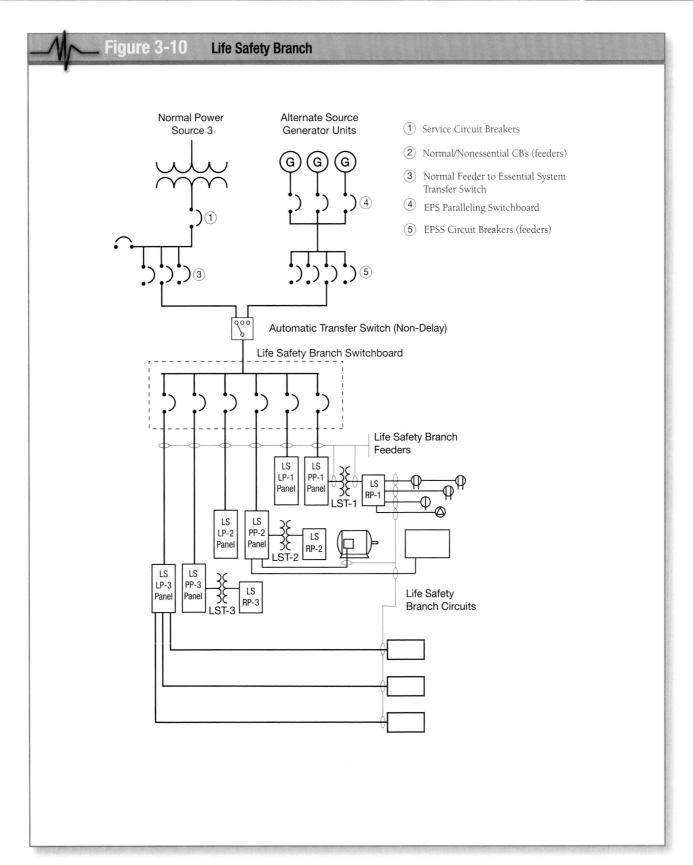

Figure 3-10. The life safety branch of the emergency system supplies power to emergency services, such as the fire alarm system.

safety system must provide a continuous supply of power for such devices in these safe areas.

The following list of items in 317.32(A) through (H) is an all inclusive list of permitted functions that need to be served by the life safety system. Unless specifically modified by building officials or under engineering supervision, additional items are not permitted to be added to this list. The following items are supplied by the life safety branch according to Section 517.32:

(A) Items providing illumination for the means of egress from the facility

(B) Exit signs

(C) Alarm and alerting systems, such as:
1. Fire alarms
2. Alarms required for systems used for the piping of nonflammable medical gases
3. Mechanical, control, and other accessories required for effective life safety systems operation

(D) Communication systems

(E) Generator set and transfer switch locations

(F) Generator set accessories

(G) Elevators

(H) Automatic doors

Three of the items above deserve further explanation. They are as follows:
1. (A) Means of egress illumination
2. (C)(3) Alarm system accessories
3. (F) Generator set accessories

Means of Egress Illumination. The means of egress in any occupancy is an important safety concern for any facility that is required to be equipped with an emergency system. Section 7.8 of *NFPA 101, Life Safety Code* provides requirements for the egress path and minimum illumination requirements. Illumination of the means of egress in a hospital is an important aspect of the emergency system, so it is logical that the life safety branch includes illumination for the means of egress. The means of egress differs from building to building in design and layout, but its purpose is essentially the same — to get persons out of the building in an emergency. The means of egress can include, but is not

limited to, corridors, stairways, passageways, landings at exit doors, and necessary ways of approach to exits.

The life safety branch of a hospital is required to provide illumination of these areas. The emergency lighting is required to be transferred to the alternate standby source within 10 seconds of a normal power failure. The lighting for patient corridors in a hospital is permitted to be switched to night lighting circuits, provided only one of two circuits can be selected and both cannot be de-energized at the same time. Once again, the concept of redundancy is present.

Alarm System Accessories. Certain HVAC controls, dampers, and certain motors are related to the safety of life. 517.32(C)(3) references mechanical controls, or other accessories that are required for effective life safety operations. Examples of such loads are smoke control and smoke evacuation systems. Even though many engineering designs and practices often specified that these systems be connected to this branch of the emergency system, previous editions of the *Code* contained no such provisions.

Generator Set Accessories. Generator set accessories include loads that are essential for continued operation of generator(s) used as an essential system alternate power source. Examples of such loads include, but are not limited to, electrically operated radiator fan motors, day tank fuel pumps, crank case heaters, etc.

Critical Branch
The *critical branch* system of the essential electrical system in a hospital is a subset of the emergency system, along with the life safety branch. It is defined as a subsystem of the emergency system consisting of feeders and branch circuits supplying energy to task illumination, special power circuits, and selected receptacles serving areas and functions related to patient care and that are connected to alternate power sources by one or more transfer switches during interruption of normal power source. [99:3.3.26] **See Figure 3-11.**

Fact

The transmission of alarms and alerts to remote panels to assist the proper authorities may appear to be a never ending list of varied equipment, from smoke and fire systems to generator and transfer equipment, etc. However, the successful transmission of these signals to the proper people is critically important. There are ways to receive, compile, and report all data received from many varied reporting systems by using a Supervisory Control and Data Acquisition (SCADA) system.

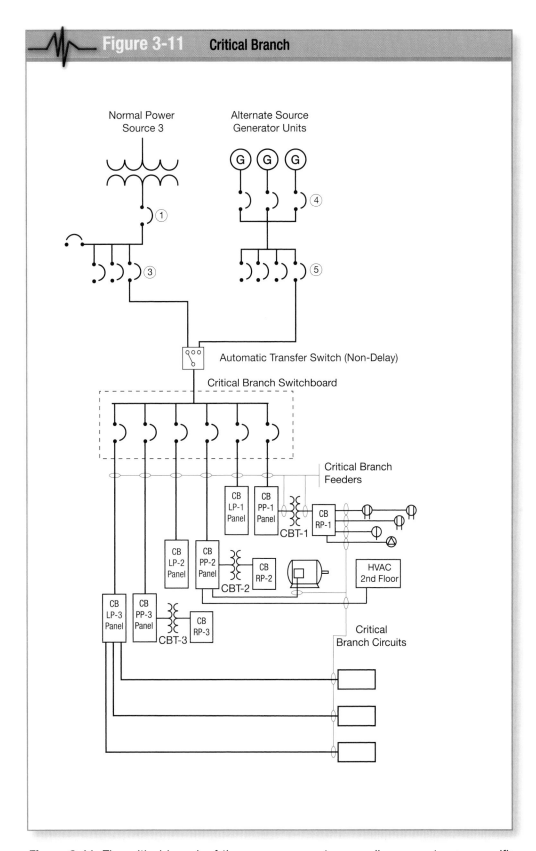

Figure 3-11. *The critical branch of the emergency system supplies power to very specific circuits related to patient care and is connected to an alternate power sources via transfer switches.*

517.19(A) includes the rules for circuit and panelboard identification for receptacles connected to the critical branch of the emergency system. The most common way to identify these receptacles is by using the color red and by including an identification marking on the faceplate that indicates which circuit supplies the receptacle and from which panelboard it is supplied. **See Figure 3-12.**

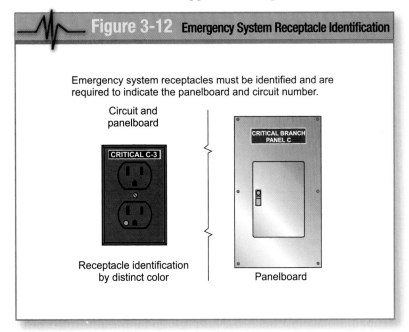

Figure 3-12 Emergency System Receptacle Identification

Emergency system receptacles must be identified and are required to indicate the panelboard and circuit number.

Circuit and panelboard

CRITICAL C-3

Receptacle identification by distinct color

CRITICAL BRANCH PANEL C

Panelboard

Figure 3-12. In addition to being identified by color, an emergency system receptacle must indicate its panelboard and circuit number.

517.33(A) requires certain specific loads to be connected to the critical care branch of a hospital emergency system. The critical branch of the emergency system in a hospital is required to supply power for task illumination, fixed equipment (such as special medical instruments and equipment), specific receptacles, and any other special power circuits that serve any of the following areas and functions related specifically to critical patient care:

1. Task illumination, selected receptacles, and fixed equipment in critical care areas that utilize anesthetizing gases
2. Isolated power systems installed in special environments

3. Task illumination and selected receptacles in any of the following patient care areas:
   - Infant nurseries
   - Medication preparation areas
   - Pharmacy dispensing areas
   - Selected acute nursing areas
   - Psychiatric areas (excluding any receptacles)
   - Ward treatment rooms
   - Nurse station illumination (unless illuminated by corridor lighting)
4. Additional specialized patient care task illumination
5. Nurse call systems
6. Blood, bone, and tissue banks
7. Telephone equipment rooms (communications equipment)
8. Task illumination for selected receptacles and power circuits for the following:
   - At least one duplex receptacle per general care patient bed location in each patient bedroom
   - Angiographic laboratories
   - Cardiac catheterization laboratories
   - Coronary care units
   - Hemodialysis rooms or areas
   - Emergency room treatment areas (selected)
   - Human physiology laboratories
   - Intensive care units
   - Postoperative recovery rooms (selected)
9. Any additional task illumination, receptacles, and selected power circuits necessary for effective operation of the hospital. 4.4.2.2.2.3(9) of *NFPA 99* also permits smaller (fractional horsepower) motors to be connected to the critical branch.

Sometimes the critical branch is subdivided into two or more branches. The *NEC* recognizes this in 517.33(B). The *Code* cautions that it is important to analyze the consequences of supplying an area with only critical branch power. It may be better to supply areas with power from both the normal and critical branch to establish and maintain desired redundancy.

NURSING HOMES AND LIMITED CARE FACILITIES

A *nursing home* is a building or portion of a building used on a 24-hour basis for the housing and nursing care of four or more persons who, because of mental or physical incapacity, might be unable to provide for their own needs and safety without the assistance of another person. [**99**:3.3.129] A nursing home is similar to a *limited care facility*, defined as a building or portion thereof used on a 24-hour basis for the housing of four or more persons who are incapable of self-preservation because of age; physical limitation due to accident or illness; or limitations such as mental retardation/developmental disability, mental illness, or chemical dependency. [**99**:3.3.97] From the electrical worker's perspective, these facilities differ from hospitals in that they do not normally offer the same level of extreme medical intervention. Therefore, the level of essential system redundancy is less extensive.

These facilities do not perform heart bypass surgeries, for instance, or organ transplants. They do not normally house people who cannot live without the support of some kind of electronic equipment. They may, however, provide services that depend on a reliable and consistent electric supply. They may offer dialysis treatments, for example. Because of the kinds of people they house, they may require a more extended service from their backup electrical system than other kinds of facilities.

If a nursing home does provide hospital care, it needs to comply with the same rules as the hospital. 517.40(B) reads, "For those nursing homes and limited care facilities that admit patients who need to be sustained by electrical life support equipment, the essential electrical system from the source to the portion of the facility where such patients are treated shall comply with the requirements of Part III, 517.30 through 517.35." **See Figure 3-13.**

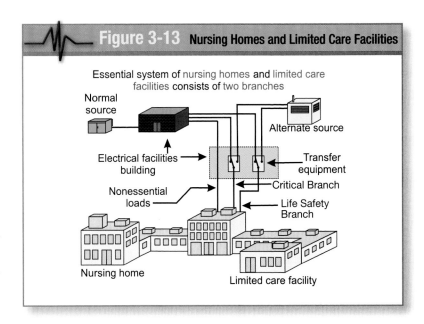

Figure 3-13. *Essential electrical systems of nursing homes and limited care facilities are made up of two branches.*

Service

A minimum of two independent power sources are required for nursing homes and limited care facilities. In this respect, nursing homes and limited care facilities resemble hospitals.

The normal source of power is usually a utility service. The alternate power source supplying the life safety and critical branches is typically an onsite generator. For rare cases where conditions require that the onsite generator serve as the normal source of power, the generator must be provided with an alternate source to supply the essential system of the facility. This alternate power source is permitted to be either another generator or an external utility service.

There is one exception that allows the alternate power source to be a battery. If the facility does not care for persons that require life support through electrical life support systems and if no surgical treatment is given that requires general anesthesia, the secondary power source may be a battery system which is either stand-alone or integrated with the individual pieces of equipment. If such a battery system is used, however, it must be effective for at least 1½ hours and capable of powering exit lights, exit corridors, stairways, nursing stations, medical preparation areas, boiler rooms, and communications areas. It must also operate all alarm systems. See *NEC* 517.40(A) Exception. Most nursing homes today are designed with an alternate source of power so that meals may be prepared and delivered to the floors and visitors may have access to patients without climbing many stairs.

The two independent electrical power sources must be independent from each other and designed to provide a high level of service continuity. 517.44(C) provides requirements for considering elements that might adversely affect the integrity of the alternate power source. The normal electrical power source is required to supply all electrical loads within the facility. The alternate power source is only required to serve as a backup, serving clearly defined loads should the normal system fail. The alternate power source must provide power for all loads on the life safety and the critical branches. There is no equipment branch of the essential system to consider, but certain

equipment within nursing homes that requires an alternate source of power is often connected to the critical branch and may be arranged for delayed connection. This will be discussed later in the critical branch section for nursing homes.

As required for hospitals, careful design considerations must be made regarding the location of the alternate power source and associated equipment used for a nursing home or limited care facility's essential electrical system and its associated components. The objective is to consider all possible causes of power interruption from reasons such as fire, storms, floods, earthquakes, or ancillary hazards created by adjoining structures or activities. It is not only important to consider causes of power interruption from foreign (outside) sources, but also the possibilities of power failures in the electrical system caused by internal facility wiring and equipment failures.

The design of an essential electrical system for these types of facilities is a huge responsibility, just as it is for hospitals. The safety and health of patients and personnel depend not only upon sound engineering design practices, but meticulous installation practices by trained electrical workers.

517.44(C) includes an important informational note that describes power sources to nursing homes and limited care facilities that are dual-fed or double-ended systems. A double-ended system is one in which two or more electrical services are fed from separate generator sets or a utility distribution network that has multiple power input sources and is arranged to provide mechanical and electrical separation so that a fault between the facility and the generating sources will not likely cause an interruption of more than one of the facility service feeders. [517.44, Informational Note] **See Figure 3-14.**

These types of arrangements are considered more reliable and provide greater degrees of service by enhancing the ability to isolate smaller portions of the power distribution system that may have failed for any reason.

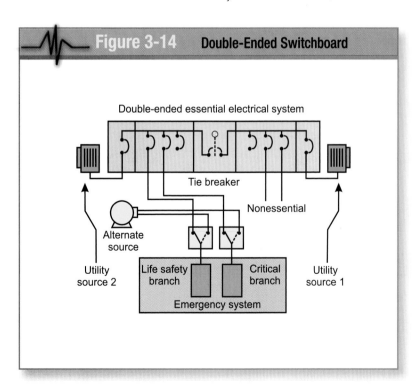

Figure 3-14 **Double-Ended Switchboard**

Double-ended essential electrical system

Tie breaker

Nonessential

Alternate source

Utility source 2

Life safety branch

Critical branch

Utility source 1

Emergency system

Figure 3-14. A double-ended system may be one with multiple feeds from a utility service combined with a local backup.

Feeder

As with hospitals, from the initial point(s) of service, or source of power, the essential electrical system feeds into a collection of transfer switches whose purpose is to transfer the branches they serve between power sources as needed.

Figure 3-14 shows at least one feeder line coming from the normal power supply into each of the transfer switches and a second line from the alternate power source into each switch. If power fails on one line, the switch will transfer its load to the other supply.

Feeders are sized based on the loads they serve, and size is calculated in accordance with the requirements for all feeders, regardless of whether they exist in hospitals or not. These requirements are covered in Articles 215 and 220 and are modified by the requirements of Article 517 (as stated in 90.3). The important thing to note in the case of health care facilities such as hospitals, nursing homes, and limited care facilities is the redundancy of both power source and feeder lines.

Branches of the Essential Electrical System

Unlike the hospital, a nursing home or limited care facility's essential electrical system consists of only two branches: life safety and critical care. **See Figure 3-15.**

Critical and life safety branches are connected to critical and life safety feeders by critical and life safety transfer switches which, in accordance with Section 700.3 of the *Code*, are required to be listed for emergency service. **See Figure 3-16.**

The purpose of transfer equipment is to recognize normal power loss, start the alternate power source, and transfer the load to the alternate power source. For an essential electrical system in a nursing home or limited care facility, this must happen within 10 seconds of the time normal power is interrupted. [517.42]

An important aspect of essential electrical system design is determining the number of transfer switches to provide in the system. Larger facilities may necessitate multiple transfer switches in their designs

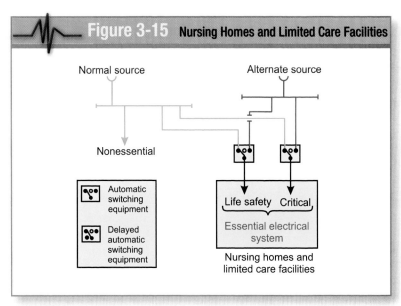

Figure 3-15. *There are two separate branches of the essential system for nursing homes and limited care facilities.*

based on the desire to maintain continuity of electric service to all areas of the facility. The minimum number of transfer switches that must be provided in a nursing home or limited care facility's essential electrical system is two, one in the critical branch and one in the life safety branch. [517.41(B)] The number of transfer switches to be used must be based on considerations of reliability, design, and

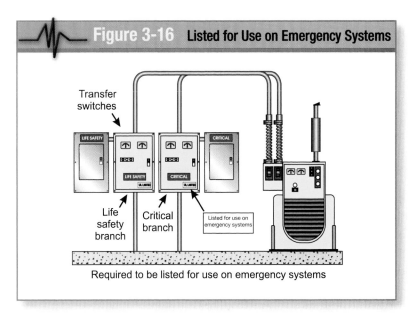

Figure 3-16. *Transfer equipment used for emergency systems is required to be listed for use on emergency systems.*

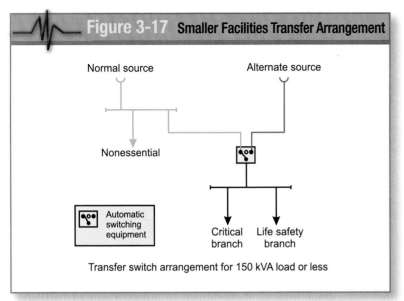

Figure 3-17   Smaller Facilities Transfer Arrangement

Normal source

Alternate source

Nonessential

Automatic switching equipment

Critical branch

Life safety branch

Transfer switch arrangement for 150 kVA load or less

Figure 3-17. Smaller facilities may use a single transfer switch, according to the Informational Note of Figure 517.41, No. 2.

connected load. The *NEC* also recognizes that for smaller facilities, installing only one transfer switch for the whole essential electrical system would be acceptable provided that maximum demand on the essential system does not exceed 150 kVA (120 KW). **See Figure 3-17.**

Essential System

Nursing homes and limited care facilities are not required to have an equipment branch in their essential electrical system configurations. However, they must have at least a life safety branch and critical care branch. Some considerations apply to both.

Circuit-Wiring Separation Requirements.

The separation requirements for essential system circuit conductors differ slightly from those in essential systems in hospitals. The life safety branch is the only branch wiring that is required to be totally independent of all other wiring, including the critical branch wiring. The life safety branch wiring is not permitted to occupy the same raceway, box, cabinet, or other enclosure with other wiring except in the following equipment:

1. Transfer equipment enclosures
2. Exit and emergency luminaires that are supplied from two sources

3. In a common junction box attached to exit or emergency luminaires that are supplied from two sources

The critical branch wiring is permitted to be installed in the same raceways, boxes, cabinets, or other enclosures with other circuit conductors that are not connected to the life safety branch. The life safety branch in a nursing home or limited care facility is at the top of the priority list, with the critical branch next, and then all other normal and equipment wiring. The reason for this priority is that it is assumed that a nursing home will not have the same need to support immobile critical care patients as a hospital, but it will need to evacuate its occupants in an emergency.

Essential System Receptacle Identification. 517.41(E) provides identification requirements for receptacles supplied by the essential system of a nursing home or limited care facility. The most common method of identification used today is the color red, but the *Code* does not specifically require red as the color to be used. For nursing homes and limited care facilities, the *Code* does not require an identification label on the faceplate that indicates which circuit supplies the receptacle and from which panelboard it is supplied. This is slightly different from the requirements for receptacles on the emergency system at a critical care patient bed location in a hospital.

Life Safety Branch

In the event of a power failure or interruption of normal service for whatever reason, the alternate power source must restore service to the life safety branch within 10 seconds after the interruption. [517.42] The life safety branch of nursing homes and limited care facilities is required to supply backup power for illumination of the required egress path in the facility. This includes all corridors, passageways, stairways, landings, exit doors, and ways of approach to these paths of egress from the building.

The *Code* permits a switching arrangement to transfer emergency lighting in patient corridors to the general illumination circuits provided only one of two circuits

The life safety system must provide backup power to illuminate exits and exit paths.

can be selected and both circuits will not be de-energized at the same time in the event of a power failure.

Sections 7.8 and 7.9 of the *NFPA 101, Life Safety Code* provide more specific details about illumination of the egress paths. The following loads are required to be connected to the life safety branch:

1. Means of egress illumination
2. Exit and directional sign illumination
3. Alarm and alerting systems such as fire alarms and gas system alarms
4. Communications systems where used for life safety instructions during emergency conditions
5. Recreation and dining area illumination (minimum levels, but sufficient)
6. Generator set loads such as task illumination and battery chargers, engine coolant fans, and so forth
7. Generator set accessories, such as radiator fan electric motors and day tank pumps
8. Elevator cab lighting, control, communications, and signaling systems [517.42(A) through (G)]
9. Automatic doors in the egress path

Critical Branch

The critical branch is the second branch of the essential system in a nursing home or limited care facility. The critical branch loads identified in 517.43(A) are required to be connected to the alternate power source, and are to be automatically connected after a short delay. The loads identified in 517.43(B) are permitted to be connected automatically or manually. The following are the types of loads that are required to be connected through automatic delay:

1. The critical branch must supply patient care task illumination and selected receptacles for medical preparation areas, pharmacy dispensing, and nurses' stations. Where the nurses' station is adequately illuminated by corridor lighting, additional lighting for the nurses' station is not required.
2. Other loads such as sump pumps are required to operate to ensure safety of associated apparatus and control systems, including alarms.
3. Loads associated with smoke control systems and stair pressurization

systems must be connected to the critical branch through a delayed automatic connection.

4. The critical branch must supply kitchen exhaust fans and supply fans that are necessary for operation if a fire should occur under the hood.

5. Any exhaust or ventilation system loads that supply airborne infectious isolation rooms are also required to be connected to the critical branch through a delayed automatic connection. [517.43(A)]

There are other loads that are permitted to be supplied by the critical branch, and arranged to be either automatically connected after a delay or manually connected after a delay. The following loads qualify for delayed automatic connection or delayed manual connection [517.43(B)]:

1. *Heating equipment for patient rooms.* The requirements for providing heat to patient rooms in nursing homes and limited care facilities may be relaxed during interruption of the normal power source where the outside design temperature is higher than 20°F. Another condition that relaxes the general requirement (even when the outside temperature is lower than 20°F) to support heating loads for patient rooms on the critical branch occurs when selected rooms are provided with heat and all confined patients can be located in them. The third alternative to providing heat for patient rooms in these types of health care facilities under power failure conditions occurs when the facility is supplied by two power sources as described in the Informational Note following 517.44(C).

2. *Elevators (throw-over provisions for temporary operation).* The provision for delayed automatic connection of elevator power is permitted where power interruption could result in patients or personnel of the facility being trapped between floors. The elevator can be

connected temporarily to the critical branch to allow the elevator to reach a landing point and release contained passengers.

3. *Lighting, receptacles, and equipment (only to the critical branch).*

DOCTORS' OFFICES, SMALL CLINICS, ETC.

The essential electrical system for a health care facility other than a hospital, nursing home, or limited care facility must be supplied by either a battery system or generator system. Where the facility does not provide life support care or critical care, the essential electrical system battery source is permitted to be in accordance with Article 700. In other words, a battery system can be used for the life safety system and must provide alternate source power for a minimum of $1^1/_2$ hours.

The essential electrical system in other health care facilities is primarily for the purpose of lighting the egress path and allowing the cessation of procedures that are ongoing, such as dental treatment or treatment being administered by doctors. The illumination can be provided by a generator that restores the minimum level of illumination necessary for the path of egress out of the building within 10 seconds as required in Section 700.12. When a dentist is providing treatment that includes drilling, cleaning, extracting, or other care associated with teeth, and the power fails, the dentist must have a minimum level of illumination to stop those procedures that are in progress. Unit equipment or battery systems can meet the minimum requirements for essential systems in such situations. 700.12(A) provides the requirements for storage battery systems and 700.12(F) provides the rules for unit equipment.

Circuit Wiring

Generally speaking, patient care areas in small clinics and doctors' office require very little "Article 517" type wiring. Except in special cases, there are no critical care or life support activities, no anesthesia other than a limited amount of conscience sedation, and no patient bed locations. There is a general

Fact

Branch circuits serving the general care areas of doctors' examining rooms and dental treatment offices require redundancy in the equipment grounding system. This is to minimize shock hazards and provide a greater degree of integrity in the effective ground-fault current path.

care area and a patient vicinity only. Because these areas are limited to administer only general care, only area receptacles and fixed equipment must meet the requirements of 517.13(A) and (B). **See Figure 3-18.**

This requirement is provided as a means of building redundancy into the equipment grounding system in areas where patients are being treated. Effective grounding will minimize shock hazards and provide a greater degree of integrity and redundancy in the effective ground fault current path should a ground fault event occur.

Storage Battery Systems

Where a storage battery is used for an essential electrical system for other health care facilities, it must have a rating and capacity suitable for supplying and maintaining the load for a period of not less than 1½ hours at not less than 87½% of the normal output capacity of the system. The batteries are required to meet the criteria for emergency service and be equipped with a suitable charging means to maintain adequate charge. There is no requirement that the batteries used for

these systems be transparent. However, where lead acid batteries that require minimum water level maintenance are used, the batteries must be transparent. Automotive batteries are not permitted to be used as this essential system power source. [700.12(A)]

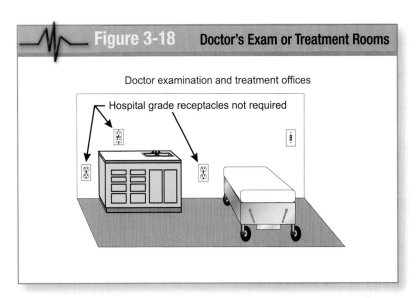

Figure 3-18 **Doctor's Exam or Treatment Rooms**

Doctor examination and treatment offices

Hospital grade receptacles not required

Figure 3-18. Branch circuits serving patient care locations such as a doctor's examining room must be wired in accordance with 517.13(A) and (B).

Transparent lead acid batteries may be used as part of the essential electrical system power source for small clinics and doctors' offices.

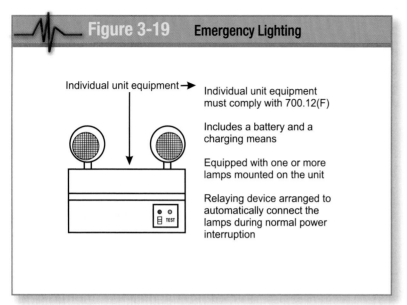

Figure 3-19. Emergency Lighting

Individual unit equipment →

Individual unit equipment must comply with 700.12(F)

Includes a battery and a charging means

Equipped with one or more lamps mounted on the unit

Relaying device arranged to automatically connect the lamps during normal power interruption

Figure 3-19. Individual unit equipment must comply with the rules of 700.12(F).

Unit Equipment Characteristics

Individual unit equipment used as the essential electrical system must be in accordance with the rules contained in 700.12(F). Unit equipment is required to consist of a battery, a charging means, provisions for one or more lamps

mounted on the equipment or remote lamps connected to the equipment, and a relaying device arranged to automatically energize the connected lamps during a normal power interruption. **See Figure 3-19.**

Unit Equipment Operation

The batteries contained within unit equipment must be of a rating suitable to provide and maintain the connected load (lamps) at not less than $87\frac{1}{2}\%$ of the nominal battery voltage for a period of time not less than $1\frac{1}{2}$ hours. The batteries are required to be designed and constructed to meet the criteria for emergency service. Unit equipment is generally required to be connected to the wiring system of the building with suitable wiring methods as provided in *NEC* Chapter 3 based on the type of construction for the building. Cord-and-plug connection is permitted where the cord and attachment plug length does not exceed 3 feet (900 mm). Any remote luminaires connected to unit equipment must be wired using one of the wiring methods in Chapter 3 of the *NEC*, such as EMT, RMC, IMC, or any listed cable wiring

A typical self-contained unit may be connected to the local area lighting branch circuit, and may have its own lamps for failover.

method other than those covered by Chapters 7 and 8. **See Figure 3-20.**

Even though the voltage supplying the remote heads may be low (12-24 VDC, for example), this wiring is part of the essential electrical system and must provide a greater degree of protection. Circuits supplying the unit equipment are required to be clearly identified in the distribution panelboard. [700.12(F)]

Battery Systems

A battery system selected for an essential electrical system must be installed to meet the provisions in Article 700. Generators installed as essential electrical systems for other health care facilities are required to meet the rules for generators in health care facilities as provided in Sections 517.30 through 517.35.

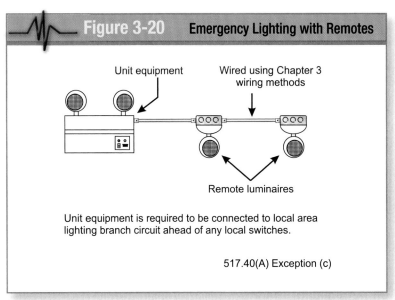

Figure 3-20 **Emergency Lighting with Remotes**

Unit equipment

Wired using Chapter 3 wiring methods

Remote luminaires

Unit equipment is required to be connected to local area lighting branch circuit ahead of any local switches.

517.40(A) Exception (c)

Figure 3-20. *Remote heads must be wired using a NEC Chapter 3 wiring method.*

Nurse Call Systems

Nurse Call Systems provide the patients and residents of health care facilities with the ability to notify staff if assistance is needed without having to leave the bed, room, or dwelling by providing a pull cord or pushbutton next to the bed which, when activated, provides both audible and visual notification to the staff.

System Overview

The following is a simple example of how a nurse call system works. A patient makes a request for staff assistance by pressing a pushbutton. The call is routed through a nurses' station unit. At the same time, a dome light outside the patient's doorway illuminates, indicating that assistance is needed and identifying the proper room from where the patient made the request. At the nurses' station unit, the call also visually appears and audibly annunciates, attracting attention and providing information on the location and status of the patient that made the call. Some nurse call systems may also annunciate at multiple locations or activate wireless pagers or phones.

A health care provider either sees the patient's dome light and/or hears the call tone. The health care provider responds directly to the call, or if at a remote station such as a nurse's desk, assigns the appropriate level of staff to answer the call. When this is done, another light above the patient's doorway may indicate that a specific level of staff is requested. The appropriate staff member(s) responds to the patient, and the call is cleared.

If the nurse call system has audio present on the system, staff-to-staff and patient-to-staff communication can take place when the call is initiated. This helps to provide the best level of service and increase staff efficiency.

Factors that can influence the type of nurse call system to be selected and installed are the type of facility, cost of the system, and which features are required. There are four major types of health care facilities that can influence the selection of a nurse call system:

1. Acute Care Facilities (Hospitals)

2. Assisted and Independent Living Facilities

3. Long Term Care Facilities (Nursing Homes)

4. Outpatient Clinics

When nurse call systems are integrated with ancillary equipment they become health care communications systems.

System Types

Tone Visual Systems. Generally speaking, Tone Visual Systems and Audio Visual Systems are two major categories of nurse call systems. The more basic form of these two nurse call systems is the Tone Visual system. Calls are generally indicated visually by a dome light over the appropriate patient's door and a light and tone on an annunciation panel or nurse master station. Call initiating devices, such as patient stations and emergency pull stations, are located at the bed, bath, and/or toilet. Patients and/or staff can signal for help using these devices.

Audio Visual Systems. An Audio Visual Nurse Call system provides the same basic patient-to-staff, staff-to-staff signaling, and annunciation functions of the Tone Visual Systems, but it is also capable of two-way audio communications. When patients or staff members place a call, it is annunciated at the nurse master station and at the patient room dome light. At this point, two-way communications can be established between the nurse and the patient/staff member who requested assistance.

Microprocessor-based Systems. Many older health care facilities have installed a conventional hard-wired logic nurse call system. These systems are rapidly being replaced with microprocessor-based systems. Both tone visual and audio visual nurse call systems can use microprocessors to enhance or replace the conventional hard-wired logic systems. Microprocessor-based system devices have a unique "address," either hardwired into the device or set via a DIP Switch installed on the device. Microprocessor-based systems can be used for various functions such as control, signaling, polling, and system supervision.

Newer microprocessor-based systems can be multiplexed, meaning that many initiating and annunciating devices can be wired together on a common system bus. This simplifies the system wiring and can reduce installation costs. This also means that multiple types of system signals can be transmitted simultaneously on the same bus cabling scheme or media. Multiplexed Systems are capable of integrating other types of equipment not always associated with the nurse call system. All devices on a multiplexed, microprocessor-based system have a unique address.

System Components

The major components of a nurse call system are the following:

1. **Nurse Master Station.** This is a device that allows call annunciation and/or verbal communication between a Nurse Desk and patients and staff members located remotely throughout the facility.

2. **Duty Station.** This is a device that provides tone and lamp annunciation to staff members based on incoming calls.

3. **Staff Station.** This is a device that is typically used as an intercom by staff to place calls to the nurse master station.

4. **Patient Station.** This is a device that allows patients and staff members to initiate calls for assistance from the patient room.

5. **Call Cord and Call Pendants.** This device is a cord connected to the Patient Station Call Socket with a single pushbutton. The patient places a call for assistance by pushing the button from their bed. Typically, a Call Pendant includes a "Call Placed" reassurance lamp to notify the patient that the nurse call system is working properly. The pendant may also include other buttons that might be used to turn on a headboard light or control speaker volume.

6. **Pillow Speakers.** This device can provide intercom communications between the patient and nurse.

7. **Call Stations and Auxiliary Devices.** The main purpose of call stations is to give staff members a way to summon help during emergencies. When pulled, the station sends a signal to annunciating equipment that flashes the appropriate lights and sound distinct tones.

8. **Dome Lamps.** This device provides visual indications for call location and status.

Wiring Requirements

Wiring requirements for nurse call systems used in health care facilities can be found in *NFPA 70*, Article 517 Health Care Facilities.

Nurse call systems shall be connected to the "Critical Branch" of the emergency power system. See *NFPA 70*, 517.33 (A)(5).

Summary

The essential electrical system within a hospital is the keystone to the wellbeing of patients and possibly their means of survival during power outages. This chapter provides a comprehensive list of the minimum electrical requirements necessary to deliver this level of performance to the facility and to the patient.

A reliable emergency power supply coupled to an adequate emergency power supply system in accordance with both *NFPA 110* and the *NEC* is the foundation of the essential electrical system. Engineers often "design in" additional features above and beyond these minimum standards in order to increase reliability, add system flexibly, and allow for future system expansion.

There are three separate levels of rules for the essential electrical system, depending on the type of occupancy. The most robust of these rules apply to hospitals. Generally, a hospital essential electrical system consists of a life safety branch, a critical branch, and one or more equipment branches. However, the normal portion of the health care electrical system shares a high level of importance as well, because it provides the major share of reliability to the overall electrical system. The second level and somewhat less robust essential electrical system applies to nursing homes and limited care facilities. Finally, the third level applies to doctors' and dental offices. Typically, this level does not provide critical care or life support.

Electrical workers must consider the enormous importance of the essential electrical system. Complying with *NEC* Chapters 1 through 4, plus Article 527 and Article 700, provides only the most minimum of electrical requirements. As a system, the essential electrical system is only as reliable as its weakest component. That component could simply be a loose locknut that delays the interruption of a single circuit breaker and causes a larger outage or interruption of power to the critical care and life support portion of the essential electrical system.

Review Questions

1. **Which branch(es) of the power distribution system in a nursing home make up the essential electrical system?**
   a. The life safety branch
   b. The critical branch
   c. The equipment branch
   d. Both a and b

2. **What are the two primary purposes of the essential electrical system in any health care facility?**
   a. Uninterrupted power for computers
   b. Uninterrupted power for surgeon's offices
   c. Continuity of power for life safety and orderly cessation of procedures
   d. None of the above

3. **The emergency system in any hospital is required to include which of the following electrical system branches?**
   a. Equipment system branch
   b. Critical branch
   c. Life safety branch
   d. Both b and c

4. **The essential electrical system of a hospital is required to include which of the following system branches?**
   a. Equipment system branch
   b. Critical branch
   c. Life safety branch
   d. All of the above

5. The essential electrical system for health care facilities is required to meet the rules in __?__ except as modified by Article 517.
    a. 500
    b. 600
    c. 700
    d. 800

6. Generally, each branch of the essential system in a hospital is required to have one or more transfer switches.
    a. True
    b. False

7. Which of the following loads are not permitted to be connected to the life safety branch of a hospital?
    a. Illumination of the means of egress
    b. Exit signs
    c. Alarm and alerting systems
    d. Kitchen hood supply and exhaust systems

8. How many power sources are required for an essential electrical system for a hospital?
    a. 1
    b. 2
    c. 3
    d. 4

9. Which of the following requirements applies to receptacles installed on the emergency system for a nursing home?
    a. They are required to be stainless steel.
    b. They are required to be red in color.
    c. The cover plates for the electrical receptacles or the electrical receptacles themselves shall have a distinctive color or marking so as to be readily identifiable as a component of the emergency system.
    d. All of the above are required.

10. One transfer switch is permitted to serve one or more branches or systems in a hospital where the maximum demand on the essential electrical system does not exceed __?__.
    a. 50 kVA
    b. 100 kVA
    c. 112.5 kVA
    d. 150 kVA

11. Mechanical protection of the emergency system in a hospital is generally required to be provided by installing the circuits in nonflexible metal raceways unless installed in any of the wiring methods provided in 517.30(C)(3)(2) through (5).
    a. True
    b. False

12. The demand calculations for a generator that supplies an essential electrical system of a hospital shall be based on which of the following?
    a. Prudent demand factors and historical data
    b. Connected loads
    c. Feeder calculation procedures described in Article 220
    d. Any combination of the above

13. Automatically operated doors in the building egress path of a hospital are required to be connected to which branch of the emergency system?
    a. Equipment branch
    b. Critical branch
    c. Life safety branch
    d. Any of the above

14. Receptacles supplied from the emergency system of the essential branch in a hospital are required to be identified by __?__.
    a. A permanent plastic label
    b. A distinctive color for the receptacle or the receptacle cover plate
    c. The color blue
    d. The color red

15. The life safety branch of a hospital is required to supply power for __?__.
    a. Lighting
    b. Receptacles
    c. Equipment
    d. All of the above

16. In a health care facility other than a hospital, nursing home, or limited care facility that provides critical care and life support equipment, the essential electrical system must include how many branches?
    a. 1
    b. 2
    c. 3
    d. 4

Inhalation Anesthetizing Locations

Until the 1968 edition of the *NEC*, Article 517 was entitled "Flammable Anesthetics." When the title of the article changed in 1971 to "Health Care Facilities," the article continued to cover flammable anesthetics under Part IV, "Inhalation Anesthetizing Locations." The title of the part clearly indicates that it covers anesthetizing locations of all kinds, not only the places where flammable agents are used. This is because many of the requirements related to flammable anesthetics are no longer relevant in the U.S. However, flammable anesthetics continue to be referenced in the *Code* because there are still places in the world where they may be an issue, and because there are older health care facilities even in the U.S. that have been wired to meet the requirements of the older safety issues caused by flammable anesthetics.

Objectives

» Explain the anesthetizing classifications of hazardous (classified) and other-than-hazardous locations

» Describe wiring methods within and above hazardous (classified) anesthetizing locations

» Describe wiring methods in other-than-hazardous anesthetizing locations

» Determine grounding and grounded power system requirements in any anesthetizing location

» Summarize the installation requirements for low voltage equipment in anesthetizing locations

» Evaluate the special issues that anesthetizing locations present for conduit sealing

Chapter 4

Table of Contents

INTRODUCTION

The early history of anesthesia included the use of several gases that posed a danger of explosion in the presence of any kind of ignition. Ignition sources ranged from obvious things such as open flames to less obvious ones like the spark that occurs when operating an everyday wall switch or simple static electricity. In 1927, for instance, an ethylene-oxygen mixture was being used for general anesthesia when static electricity caused an explosion that rocked the four-story building, killing several people. In the 1930s, the American Medical Association (AMA) concluded that explosion hazards in anesthesia processes were a widespread concern.

In the 1950s, developments in chlorofluorocarbon chemistry produced halogenated, nonflammable agents that replaced the explosive agents. More than 20 years ago the Joint Commission on Accreditation of Hospitals (JCAH) in its Accreditation Manual for Hospitals prohibited the use of flammable anesthetic agents in all anesthetizing locations. An *anesthetizing location* is any area of a facility that has been designated to be used for the administration of any flammable or nonflammable inhalation anesthetic agent in the course of examination or treatment, including the use of such agents for relative analgesia. *Relative analgesia* is a state of sedation and partial block of pain perception produced in a patient by the

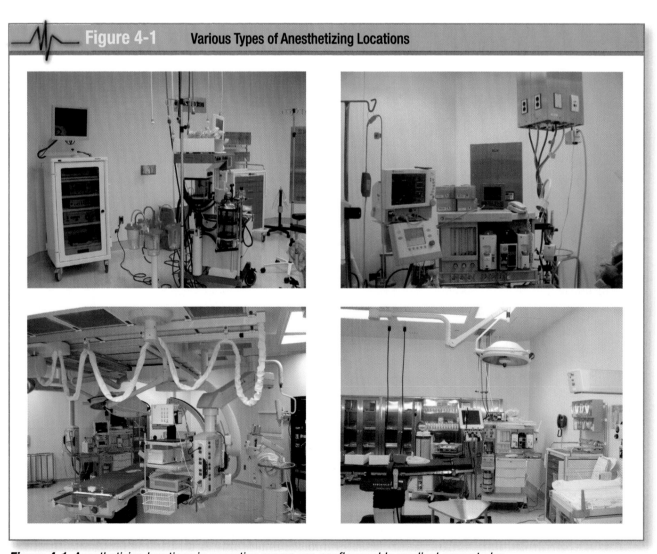

Figure 4-1 Various Types of Anesthetizing Locations

Figure 4-1. *Anesthetizing locations in operating rooms use nonflammable medical gases today.*

The electrical requirements in the *Code* for health care facilities include many electrical system rules that are also part of *NFPA 99*. Past editions of *NFPA 99, Standard for Health Care Facilities*, contained requirements for flammable anesthetizing locations. A *flammable anesthetizing area* is any area of a facility that has been designated to be used for the administration of any flammable inhalation anesthetic agents in the normal course of examination or treatment. For the most part, these do not exist anymore. These requirements have been relocated to Annex E of *NFPA 99*, however, and the reason is explained in the note to Annex E, which reads as follows:

> *NOTE: The text of this annex is a compilation of requirements included in previous editions of NFPA 99 on safety practices for facilities that used flammable inhalation anesthetics. This material is being retained in this annex by the Technical Committee on Anesthesia Services for the following reasons: (1) The Committee is aware that some countries outside the United States still use this type of anesthetics and rely on the safety measures herein; and (2) while the Committee is unaware of any medical schools in the United States still teaching the proper use of flammable anesthetics or any health care facilities in the United States using flammable anesthetics, retaining this material will serve as a reminder of the precautions that would be necessary should the use of this type of anesthetics be reinstituted.* [**99**:Annex E]

> *(Excerpt from NFPA 99.)*

Operating rooms of today, the primary locations for anesthesia use, have other electrical dangers, however, that make the continued use of such protective measures valuable even in the absence of flammable anesthetics. For instance, many operating rooms were classified as wet locations according to previous editions of the *NEC*. However, for the 2008 *Code*, the use of wet location has been changed to wet procedure locations, and therefore require special protection techniques to prevent electrical shock injuries.

inhalation of concentrations of nitrous oxide insufficient to produce loss of consciousness (conscious sedation).

These developments greatly reduced the risk of fire and explosion previously present in all anesthetizing locations. **See Figure 4-1.**

Today, explosion hazards are statistically the least of the hazards associated with anesthesia. Post-operative respiratory complications and pneumonia are greater concerns related to anesthesia use than the dangers of explosion. However, the *NEC* continues to cover proper wiring and grounding techniques for areas that have been designated as a hazardous (classified) location based on the presence of flammable anesthetics because there are still areas of the world where

such anesthetics are or may be used, and also because in the U.S. there are legacy institutions that have been designed to the standards of an earlier age. *Flammable anesthetics* include gases or vapors, such as fluroxene, cyclopropane, divinyl ether, ethyl chloride, ethyl ether, and ethylene, which may form flammable or explosive mixtures with air, oxygen, or reducing gases such as nitrous oxide. Today's electrical workers still need to understand the requirements of such places when they are asked to work in them.

CLASSIFICATION

The general requirements of area classification are found in all fire codes and building codes throughout the world. Basically these codes require that any flammable liquid, gas, or vapor that could be present within or near buildings be used, treated, and worked with in a safe manner to prevent ignition, fire, explosion, and injury to persons and property. The electrical requirements for safely using electrical equipment or placing any wiring in close proximity to these hazardous (classified) locations is covered in Chapter 5, Articles 500 and 501. Specifically, Section 500.4 begins the special requirement process. Anesthetizing locations may be considered hazardous or other-than-hazardous; and either of these may or may not also be considered a wet procedure location. The terms *hazardous* and *other-than-hazardous* are used in a very narrowly defined sense. In the sense used by the *Code* in dealing with anesthetizing locations, the term *hazardous* specifically refers to a location where flammable anesthetics are present. An area may still be a dangerous area and not be classified "hazardous" in this sense if the danger is not the result of flammable anesthetics. For this reason, the term *other-*

than-hazardous is used for other areas which may or may not be dangerous, but because they do not involve flammable anesthetics, are not classified as "hazardous" in this very technical use of the term.

Wet Procedure Locations

If an anesthetizing location is also considered a wet procedure location, all the rules for wet procedure locations apply in addition to any other rules that may apply. Special requirements applicable to wet procedure locations in patient care areas of health care facilities are covered by the *Code* in 517.20(A) and (B).

Hazardous Locations

According to the *NEC*, there are different types of hazardous locations. For anesthetizing locations, the hazard is one which is created by the presence of flammable gases or vapors in the air. When these materials are found in the atmosphere, a potential for explosion exists, which could be ignited if an electrical or other source of ignition is present. The *Code* refers to this first type of hazard as Class I. A Class I Hazardous Location is one in which flammable gases or vapors may be present in the air in sufficient quantities to be explosive or ignitable.

In addition to the type of hazardous location, the *Code* also considers the kinds of conditions under which these hazards are present. The *Code* specifies that hazardous material may exist in several different kinds of conditions which can be described as normal conditions or abnormal conditions.

In the normal condition, the hazard would be expected to be present in everyday production operations or during frequent repair and maintenance activity. When the hazardous material is expected to be confined within closed containers or closed systems and will be present only through accidental rupture, breakage, or unusual faulty operation, the situation could be called "abnormal." The *Code* designates these two kinds of conditions as Division 1 (normal) and Division 2 (abnormal). Therefore, Class I hazardous locations can be either Division 1 or Division 2.

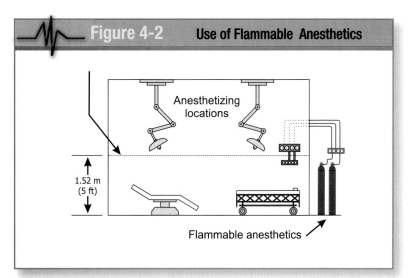

Figure 4-2 **Use of Flammable Anesthetics**

Anesthetizing locations

1.52 m (5 ft)

Flammable anesthetics

Figure 4-2. Where flammable anesthetizing agents are employed, the entire area of the room extending up to 5 feet (1.52 m) above the finished floor is a Class I, Division 1 location.

In the past, when flammable anesthetics were used in the U.S., an operating room using such anesthetics would have been classified a Class I, Division 1 location because it is not unusual for flammable or explosive vapors to be present outside of closed containers or piping during a normal procedure.

Hazardous (classified) anesthetizing locations do not exist in the United States anymore because flammable anesthesia is no longer used. Hospital personnel are no longer trained in how to work in such environments. In fact, they are told that if they ever have to work in such an environment – in a foreign country or under emergency conditions, perhaps – they should perform their operations outdoors so the hazardous vapors can escape before creating an explosive atmosphere. However, because an electrical worker may still be required to work in a hazardous (classified) location (in a foreign country or a legacy building in the U.S.), the *Code* continues to cover the requirements for such locations.

Area of use
In an area where flammable anesthetics are used, the entire area is considered a Class I, Division 1 location extending from the lowest point in the room or area upward to a level of 5 feet (1.52 m) above the floor. The remaining volume up to the structural ceiling is considered to be above a hazardous location and also has special electrical requirements. [**99:** Annex E, E.1, and E.2] 517.60(A)(1) does not provide a Class I, Division 2 (Division 2 is less restrictive) location adjacent to the defined Division 1 boundary. Specific information and rules for wiring and equipment installed above the hazardous location are provided later in this chapter. The governing body of the

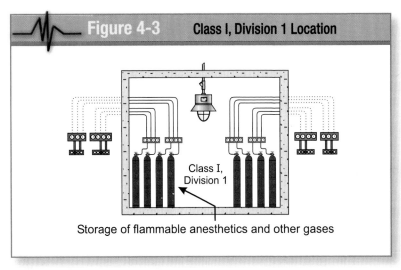

Figure 4-3. *Flammable anesthetizing agent and disinfectant storage rooms are Class I, Division 1 locations.*

facility is responsible for designating these locations and provides information about what type anesthetizing agents are to be used within the facility for determining the extent of the Class I, Division 1 location boundaries. **See Figure 4-2.**

Area of Storage
In a health care facility, any room or location in which flammable anesthetics, or volatile flammable agents of any kind, are stored is considered a Class I, Division 1 location from floor to ceiling. **See Figure 4-3.**

Other-than-Hazardous Locations
Any inhalation anesthetizing location designated for the exclusive use of nonflammable anesthetizing agents is considered an other-than-hazardous location. The term is "other-than-hazardous" as opposed to "non-hazardous" because the *Code* uses the term "hazardous" in a very strict sense. The term refers to areas that are specifically classified as hazardous (classified) locations in accordance with Article 500 because fire and explosion hazards may occur where flammable anesthetics are involved. There are other areas that may be unsafe for reasons that have nothing to do with anesthetic use. They are, however, "other-than-hazardous" locations because any hazard they may have is not related to the use of flammable anesthetics.

Fact

500.8 (C) indicates that equipment in a hazardous (classified) location must be marked to show the environment for which it has been evaluated, such as class, division, group, temperature, ambient, and any special allowances.

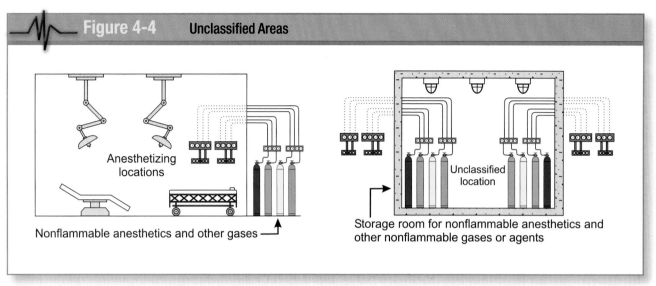

Figure 4-4   **Unclassified Areas**

Anesthetizing locations

Nonflammable anesthetics and other gases

Unclassified location

Storage room for nonflammable anesthetics and other nonflammable gases or agents

Figure 4-4. Operating rooms where only nonflammable anesthetics and disinfecting agents are employed are unclassified (that is, other-than-hazardous) locations.

Where there are no flammable anesthetizing agents, flammable disinfecting agents, or flammable laboratory gases used, there is no requirement for hazardous location classification. These locations are therefore unclassified and general wiring methods and equipment suitable for an anesthetizing location are permitted to be used. **See Figure 4-4.**

Where there are no flammable anesthetizing agents, flammable disinfecting agents, or flammable laboratory gases stored, there is no requirement for

hazardous location classification as well. These locations are therefore unclassified, and again general wiring methods as modified by Article 517 are permitted to be used. **See Figure 4-5.**

WIRING AND EQUIPMENT
Wiring and equipment requirements for anesthetizing areas vary depending on whether an area is considered a hazardous location, above a hazardous location, or an other-than-hazardous location.

Figure 4-5   **Medical Gas Storage Rooms**

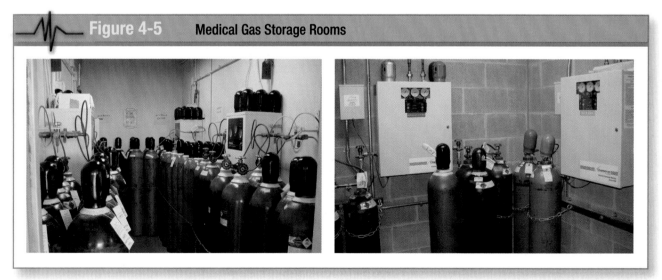

Figure 4-5. Storage rooms for nonflammable anesthetizing agents, disinfectants, and reducing gases are unclassified locations.

Within Hazardous Locations

The following requirements apply to wiring and equipment within a hazardous location:

- **Isolation.** Except as permitted in 517.160 (which covers isolated power systems, discussed below and in other chapters of this book), each power circuit within, or partially within, a flammable anesthetizing location must be isolated from any distribution system by the use of an isolated power system. [**99:** Annex E, E.6.6.2] **See Figure 4-6.**

- **Design and installation.** Where an isolated power system is utilized, the isolated power equipment must be listed as isolated power equipment, and the isolated power system must be designed and installed in accordance with 517.160 and its listing and labeling. Isolated power systems are covered in more detail in other chapters of this book.

- **Equipment operating over 10 volts.** In hazardous locations, all fixed wiring and equipment and all portable equipment, including lamps and other utilization equipment, operating at more than 10 volts between conductors must comply with the requirements of 501.1 through 501.25, 501.30(A) and 501.30(B), and 501.100 through 501.150. These articles cover the requirements for Class I, Division 1 locations. All equipment operating over 10 volts, used within a hazardous location of a health care facility, must be specifically approved for the hazardous atmospheres involved. [**99:**Annex E, E.2.1, E.4.5, E.4.6, and E.4.7]

- **Extent of location.** Where a box, fitting, or enclosure is partially, but not entirely, within a hazardous location, the hazardous location is extended to include the entire box, fitting, or enclosure.

- **Receptacles and attachment plugs.** Receptacles and attachment plugs in a hazardous location must be listed for use in Class I, Group C hazardous locations and must have

Fact

Isolated power systems provide protection not only from explosion in the presence of flammable gases, but also from line-to-ground shock under wet conditions, and from power loss during a ground-fault condition in critical care situations. All of these situations apply in a wet procedure location.

provision for the connection of a grounding conductor.

- **Flexible cord type.** Flexible cords used in hazardous locations for connection to portable utilization equipment, including lamps operating at more than 8 volts between conductors, must be of a type approved for extra hard usage in accordance with Table 400.4 of the *NEC* and must include an additional conductor for grounding.

- **Flexible cord storage.** A storage device for the flexible cord must be provided and must not subject the cord to bending at a radius of less than 3 inches (75 mm). The common method of meeting this cord storage requirement is to use cord storage reels mounted on the ceiling. Because flammable anesthetizing agents are not used any more, this requirement is not often found in health care facilities.

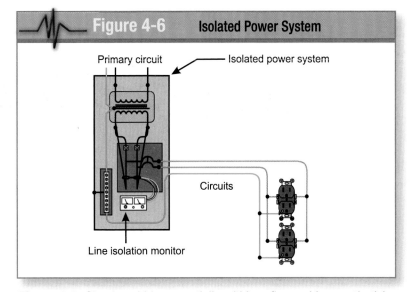

Figure 4-6 **Isolated Power System**

Primary circuit — Isolated power system

Circuits

Line isolation monitor

Figure 4-6. *Circuits within or partially within a flammable anesthetizing location must be supplied from an isolated power system.*

Above Hazardous Locations

The following requirements apply to wiring and equipment above hazardous locations:

- **Wiring methods.** Wiring above a hazardous location must be installed in rigid metal conduit, electrical metallic tubing, intermediate metal conduit, Type MI cable, or Type MC cable that employs a continuous, gas/vapor-tight metal sheath.

- **Equipment enclosure.** Installed equipment that may produce arcs, sparks, or particles of hot metal, such as lamps and lamp holders for fixed lighting, cutouts, switches, generators, motors, or other equipment having make-and-break or sliding contacts, must be of the totally enclosed type or be constructed so as to prevent escape of sparks or hot metal particles.

  There is an exception to this in the *Code*. Wall-mounted receptacles installed above hazardous locations are not required to be totally enclosed or have their openings screened. This would defeat the purpose of the receptacle. Receptacles do have their own special requirements which are listed separately below.

- **Luminaires.** Surgical and other luminaires must conform to 501.130(B) which covers Class I,

Division 2 locations, with the following two exceptions:

1. Surface temperature limits set in 501.130(B)(1) do not apply to these luminaires.
2. Integral or pendant switches that are located above and cannot be lowered into the hazardous location need not be explosionproof.

- **Seals.** Listed seals must be provided in conformance with 501.15 and 501.15(A)(4), which cover sealing and drainage requirements for Class I, Division 1 locations. These same sections of *Code* apply to horizontal as well as to vertical boundaries of the defined hazardous location. Conduit seals are discussed in more detail later in this chapter.

- **Receptacles and attachment plugs.** Receptacles and attachment plugs located above hazardous anesthetizing locations must be listed for hospital use for services of prescribed voltage, frequency, rating, and number of conductors with provision for the connection of the grounding conductor. This requirement applies to attachment plugs and receptacles of the 2-pole, 3-wire grounding type for single-phase, 120-volt, nominal, AC service. **See Figure 4-7.**

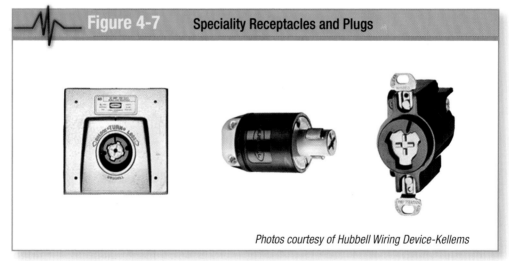

Figure 4-7   Speciality Receptacles and Plugs

Photos courtesy of Hubbell Wiring Device-Kellems

Figure 4-7. *Special receptacles listed for hospital use allow only mated attachment plugs to be inserted.*

- *250-volt receptacles and attachment plugs rated 50 and 60 amperes.* Receptacles and attachment plugs rated 250 volts, for connection of 50-ampere and 60-ampere AC medical equipment for use above hazardous locations, must be arranged so that the 60-ampere receptacle will accept either the 50-ampere or the 60-ampere plug. Fifty-ampere receptacles must be designed so as not to accept the 60-ampere attachment plug. The attachment plugs must be of the 2-pole, 3-wire design with a third contact connecting to the insulated (green or green with yellow stripe) equipment grounding conductor of the electrical system.

Other-Than-Hazardous (Classified) Locations

The following requirements apply to wiring and equipment in other-than-hazardous (classified locations:

- **Wiring methods.** Wiring serving other-than-hazardous locations, as defined in 517.60, must be installed in a metal raceway system or cable assembly. The metal raceway system or cable armor or sheath assembly must qualify as an equipment grounding conductor in accordance with 250.118. Type MC and Type MI cable must have an outer metal armor, sheath, or sheath assembly that is identified as an acceptable equipment grounding conductor.

  There is an exception to this. Pendant receptacle installations that employ listed Type SJO or equivalent hard usage or extra-hard usage, flexible cords suspended not less than 6 feet (1.8 m) from the floor need not be installed in a metal raceway or cable assembly.

- *Receptacles and attachment plugs.* Receptacles and attachment plugs installed and used in other-than-hazardous locations must be listed "hospital grade" for the prescribed voltage, frequency, rating, and number of conductors with provision for connection of the grounding

> **Fact**
>
> According to 517.62(C), receptacles and plugs installed and used in other-than-hazardous (classified) locations must be listed "hospital grade" for services of prescribed voltage, frequency, rating, and number of conductors.

conductor. This requirement must apply to 2-pole, 3-wire grounding type for single-phase, 120-, 208-, or 240-volt, nominal, AC service.

- *250-volt receptacles and attachment plugs rated 50 and 60 amperes.* Receptacles and attachment plugs rated 250 volts, for connection of 50-ampere and 60-ampere AC medical equipment for use in other-than-hazardous locations, must be arranged so that the 60-ampere receptacle will accept either the 50-ampere or the 60-ampere plug. Fifty-ampere receptacles must be designed so as not to accept the 60-ampere attachment plug. The attachment plugs must be of the 2-pole, 3-wire design with a third contact connecting to the insulated (green or green with yellow stripe) equipment grounding conductor of the electrical system.

GROUNDING AND BONDING IN HAZARDOUS (CLASSIFIED) LOCATIONS

In any anesthetizing area, all metal raceways and metal-sheathed cables and all normally non-current-carrying conductive portions of fixed electrical equipment (unless the equipment operates at no more than 10 volts between conductors) must be connected to an equipment grounding conductor. Grounding and bonding requirements must conform to the rules of Section 501.30, which cover grounding and bonding requirements for Class I locations (Division 1 and Division 2). This more restrictive method of bonding must extend all the way back to the point of grounding for the derived system or service.

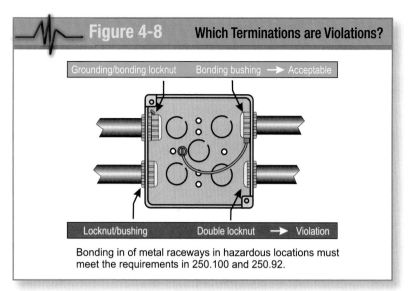

Figure 4-8 **Which Terminations are Violations?**

Grounding/bonding locknut   Bonding bushing → Acceptable

Locknut/bushing   Double locknut → Violation

Bonding in of metal raceways in hazardous locations must meet the requirements in 250.100 and 250.92.

Figure 4-8. Both acceptable and unacceptable bonding means for hazardous locations wiring are shown here.

The *Code* places some special requirements for grounding and bonding conductive equipment in hazardous locations. These requirements can be found in 501.30(A) and 250.100 for Class I locations. **See Figure 4-8.**

In hazardous locations it is vital to have effective grounding and bonding to prevent an explosion. Under fault-current conditions, when heavy currents are present in metal conduit, every

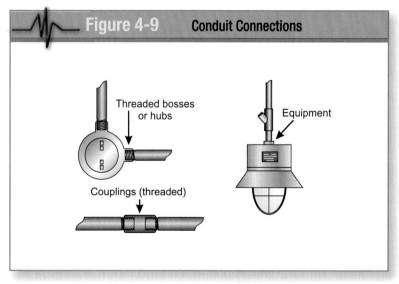

Figure 4-9 **Conduit Connections**

Threaded bosses or hubs

Equipment

Couplings (threaded)

Figure 4-9. Type NPT threaded entries into explosionproof equipment must be made wrench-tight and generally with at least five threads fully engaged.

connection point in the raceway system is a potential source of sparks and ignition. If there is an arcing fault to a metal enclosure in a hazardous location, the external surface temperature of the metal enclosure at the point of the arcing fault will start to rise to temperatures that could cause ignition of the flammable vapors. Under these fault conditions, it is essential that the overcurrent device operate as quickly as possible to prevent a hot spot on the enclosure or arcs that may burn through the enclosure from igniting the atmosphere on the outside of the enclosure. It is extremely important that all threaded joints be made wrench-tight to prevent sparking at those threaded joints. If joints are other than the threaded type, such as locknuts and bushings or double locknuts and bushings at boxes, enclosures, cabinets, and panelboards, it is essential that bonding be assured around those joints in the fault-current bonding path to prevent sparking and assure a low-impedance path for the fault current. **See Figure 4-9.**

General bonding requirements are found in Part V of Article 250 of the *NEC*. Section 250.90 requires bonding "where necessary to ensure electrical continuity and the capacity to conduct safely any fault current likely to be imposed." Section 250.100 includes bonding requirements for hazardous locations and indicates that, regardless of the voltage, the electrical continuity of non-current-carrying metal parts of equipment, raceways, and other enclosures in any hazardous location shall be ensured by any of the methods specified for services in 250.92(B) that are approved for the wiring method used.

During low impedance ground-fault conditions there are heavy currents flowing through metal conduits in the effective ground-fault current path. Every joint or connection point, such as couplings, locknuts, hubs, and other fittings, can be a potential source of ignition from arcing or sparks. It is imperative that all threaded fittings and joints be made up wrench-tight to prevent arcing from these threaded joints for the duration of time it takes the overcurrent device to clear the fault. Fittings suitable for the bonding required must be used. The types of methods acceptable to meet these requirements are provided in 250.92(B).

Generally, locknuts on each side of the enclosure, or a locknut on the outside and a bushing on the inside, cannot be used for bonding. Bonding locknuts or bonding bushings with bonding jumpers must be used to ensure the integrity of the bond and its ability to carry the fault current that may be imposed without arcing or sparking at the connections.

The bonding means must generally be installed from the hazardous location to the service equipment or point of grounding of any separately derived system that is the source of the circuit. This includes all raceways, fittings, junction boxes, enclosures, controllers, and panelboards between the hazardous location and the service or separately derived system. The goal is to ensure a low impedance path for ground-fault currents and to facilitate fast operation of overcurrent protective devices supplying the circuit in the hazardous locations. **See Figure 4-10.**

GROUNDED POWER SYSTEMS

Grounded power systems in anesthetizing locations have requirements that vary depending on the type of power system and whether the location is classified as hazardous, above a hazardous location, or other-than-hazardous. Except as noted, the requirements below apply to any anesthetizing location.

Battery-Powered Emergency Lighting

One or more battery-powered emergency lighting units must be provided in accordance with 700.12(F). These units are allowed to be connected to the critical branch lighting circuit and connected ahead of any local switching. Such lighting units consist of:

- A rechargeable battery
- A battery charging means
- Provisions for one or more lamps mounted on the equipment, or with terminals for remote lamps, or both
- A relaying device arranged to energize the lamps automatically upon

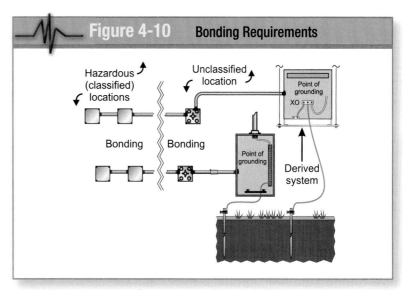

Figure 4-10. Bonding Requirements

Figure 4-10. *Bonding requirements must extend to all intervening metal raceways all the way back to the applicable service or source of separately derived system grounding point.*

failure of the supply to the unit equipment

The batteries must have the capacity to operate for a period of at least $1\frac{1}{2}$ hours at $87\frac{1}{2}\%$ of the nominal battery voltage for the total lamp load associated with the unit; or the unit must be able to supply at least 60% of the initial emergency illumination for a period of at least $1\frac{1}{2}$ hours.

Unit equipment must be permanently fixed in place (that is, not portable) and have all wiring to each unit installed in accordance with the requirements of any of the applicable wiring methods permitted by Parts III and IV of Article 517.

The branch circuit feeding the unit equipment must be the same branch circuit as that serving the normal lighting (or critical lighting as permitted) in the area and connected ahead of any local switches. There is an exception to this rule if the area is supplied by a minimum of three normal lighting circuits. In this case, a separate branch circuit for the emergency lighting units is allowed as long as it originates from the same panelboard as the other lighting circuits and is provided with a lock-on feature. In either case, the branch circuit that feeds unit equipment must be clearly identified at the distribution panel.

For additional information, visit qr.njatcdb.org Item #1054

Fact

It is important to note how "Unit Equipment" and its application in 700.12(F) differs from "Battery-Powered Lighting Units" as defined in 517.2 and applied in the 517.63(A) provisions of Part IV, Inhalation Anesthetizing Locations, differ.

Branch-Circuit Wiring

Branch circuits supplying only listed, fixed, therapeutic, and diagnostic equipment, permanently installed above a hazardous location and in other-than-hazardous locations, are allowed to be supplied from a normal grounded service, single- or three-phase system, provided the following apply:

- Wiring for grounded and isolated circuits does not occupy the same raceway or cable.
- All conductive surfaces of the equipment are connected to an equipment grounding conductor.
- Equipment (except enclosed X-ray tubes and the leads to the tubes) is located at least 8 feet (2.5 m) above the floor or outside the anesthetizing location. This applies only to hazardous locations, not other-than-hazardous ones.
- Switches for the grounded branch circuit are located outside the hazardous location. This also only applies to hazardous locations, not other-than-hazardous ones.

Fixed Lighting Branch Circuits

Branch circuits supplying only fixed lighting are permitted to be supplied by a normal grounded service under the following conditions:

- Such luminaires are located at least 8 feet (2.5 m) above the floor. This applies only to hazardous locations, not other-than-hazardous ones.
- All conductive surfaces of the luminaires are connected to an equipment grounding conductor.
- Wiring for circuits supplying power to luminaires does not occupy the same raceway or cable for circuits supplying isolated power.

Fact

Properly maintained battery-powered lighting units supplied from the critical branch are the most reliable source of light in an anesthetizing location because they are supplied by three distinct sources of power.

- Switches are wall-mounted and located above hazardous locations. This applies only to hazardous locations, not other-than-hazardous ones.

Remote-Control Stations

Wall-mounted remote-control stations for remote-control switches operating at 24 volts or less are allowed in any anesthetizing location.

Isolated Power System Location

Where an isolated power system is used, the isolated power equipment must be listed as isolated power equipment. Isolated power system equipment and its supply circuit are allowed to be located in an anesthetizing location, provided they are installed above a hazardous location or located in an other-than-hazardous location.

Circuits

Circuits in anesthetizing locations within or partially within the Division I area of the hazardous (classified) location must be isolated from any distribution system supplying other-than-anesthetizing locations, except of course, the supply circuit to an isolated power system located within the anesthetizing location.

LOW-VOLTAGE EQUIPMENT AND INSTRUMENTS

Low-voltage equipment is used in all anesthetizing locations. In hazardous locations, even low voltage equipment can present a danger. Any interruption of the circuit, even circuits as low as 10 volts, by switch or loose or defective connection anywhere in the circuit, may produce a spark sufficient to ignite flammable anesthetic agents.

Equipment Requirements

Low-voltage equipment that is frequently in contact with the bodies of persons or has exposed current-carrying elements must comply with one of the following:

- Operate on an electrical potential of 10 volts or less
- Be approved as intrinsically safe or double-insulated equipment
- Be moisture resistant

Fact

Low voltage equipment and instruments used in an other-than-hazardous anesthetizing location must meet the general requirements of Sections 90.7, 110.2, and 110.3. In addition, all low-voltage equipment must also meet the requirements of 517.64.

Power Supplies

Power for low-voltage equipment must be supplied by one of the following:

- An individual portable isolating transformer (not an autotransformer) connected to an isolated power circuit receptacle by means of an appropriate cord and attachment plug
- A common low-voltage isolating transformer installed in an other-than-hazardous location
- Individual dry-cell batteries
- Common batteries made up of storage cells located in an other-than-hazardous location

Isolated Circuits

Isolating-type transformers for supplying low-voltage circuits must have both of the following:

- Approved means for insulating the secondary circuit from the primary circuit
- The core and case connected to an equipment grounding conductor

Controls

Resistance or impedance devices are allowed to control low-voltage equipment, but only certain power-limited power supplies using isolation techniques or batteries are allowed, since the maximum voltage to the equipment must be limited by a reliable method.

Battery-Powered Appliances

Battery-powered appliances must not be able to charge while in operation unless their charging circuitry incorporates an integral isolating-type transformer.

Receptacles or Attachment Plugs

Any receptacle or attachment plug used on low-voltage circuits must be of a type that does not permit interchangeable connection with circuits of higher voltage.

OTHER CONSIDERATIONS

In addition to the requirements of Article 517.60 specifically related to inhalation anesthetizing locations themselves, there are requirements for other areas of a health care facility that are the direct result of having anesthetizing locations on their premises. The piping that the gas flows through requires special consideration from the electrical worker's perspective. Also, because of the piping, there are issues around sealing the spaces pipes must penetrate as well as the spaces electrical conduit must pass through when passing through the boundaries of an anesthetizing location.

Gas Piping

Anesthetizing gases are connected to the manifold of the gas piping and delivery system of the facility from their storage locations. **See Figure 4-11.**

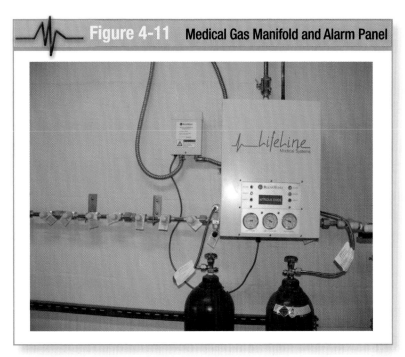
Figure 4-11   **Medical Gas Manifold and Alarm Panel**

Figure 4-11. *Nonflammable anesthetizing gas bottles are connected to the manifold and piping system of a health care facility.*

Figure 4-12 **Fire-Stopped Wall Penetrations**

Figure 4-12. *Nonflammable anesthetizing gas system piping must be sealed properly.*

The anesthetizing gas systems are sealed systems and piped throughout the facility in accordance with applicable requirements from gas system standards. As medical gas piping passes through fire-rated floors and walls, those penetrations must be sealed to maintain the initial fire rating of the wall or floor assembly. **See Figure 4-12.**

Also, single and multiple medical gas piping outlets are installed using fire-rated outlets or cabinets that retain the same fire-rating as that of the wall assembly. **See Figure 4-13.**

According to Article 250 Grounding and Bonding, metal gas piping systems in health care facilities fall under the category of other metal piping systems that are

Figure 4-13 **Medical Gas Outlet Panel**

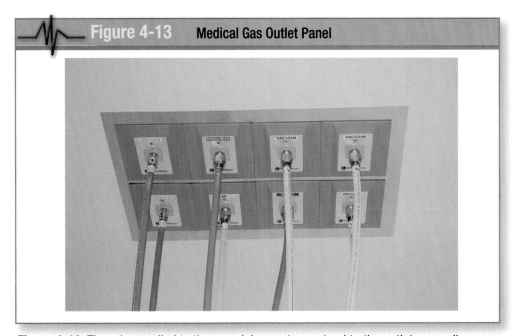

Figure 4-13. *The rules applied to the gas piping system extend to the outlets as well.*

Figure 4-14 Mechanical Piping Systems

Figure 4-14. Other metal piping systems in health care facilities are required to be electrically bonded in accordance with 250.104(B).

required to be bonded to the electrical power wiring system in accordance with 250.104(B). This rule requires other metal piping, including metal gas piping likely to be energized, to be bonded to the service equipment enclosure, the grounded conductor at the service, the grounding electrode conductor, where large enough, or to one or more of the grounding electrodes used for the service. By the minimum requirements in 250.104(B), the equipment grounding conductor of the circuit likely to energize the gas piping or other piping system is permitted as the bonding means. The minimum size of the metal gas piping system bonding jumper is required to be not less than the sizes provided in 250.122 using the rating of the overcurrent device of the branch circuit likely to energize the metal piping system. **See Figure 4-14.**

Conduit Sealing

Hospitals are typically built using Type 1 construction (a fire-resistive type of construction as clarified in *NFPA 99*, Annex E); therefore, the wiring methods used are usually metallic conduit, tubing, or listed metallic cable assemblies. As stated before, flammable gases can, over time, migrate into electrical enclosures, and when an explosive mixture is present and an electrical spark is present, the air/gas mixture can ignite. Explosionproof enclosures together with sealed conduit systems are installed in a manner that prevents these minor ignitions from spreading into the general environment and causing explosions or fires. Directly attaching conduit seals to explosionproof enclosures actually "completes and seals" the enclosure.

Therefore, these seals are intended to prevent the passage of flames from one

 Fact

Seals are provided in conduit systems to minimize the passage of gases and vapors and to prevent the passage of flames from one portion of the electrical installation to another through the conduit. Seals in conduit must comply 501.15(A) through (F).

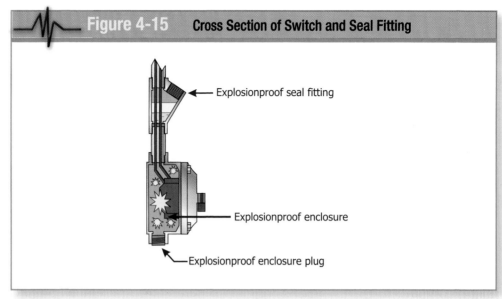

Figure 4-15. *Conduit seal fittings are required to maintain explosionproof integrity of equipment.*

portion of the electrical installation to another. Conduit and cable seals are required in hazardous locations and serve two primary purposes: (1) to contain explosions to explosionproof enclosures, and (2) to prevent the exploding gases and flames from migrating into the conduit system during such events. **See Figure 4-15.**

Sealing is also required to minimize the passage or communication of gases or vapors from a Class I location to one that is unclassified. Suitable conduit and cable seals are to be provided at horizontal and vertical boundaries of the anesthetizing locations defined as hazardous. **See Figure 4-16.**

Seals are required to meet the applicable rules in 501.15. The rules for conduit seals apply to conduit seals installed in

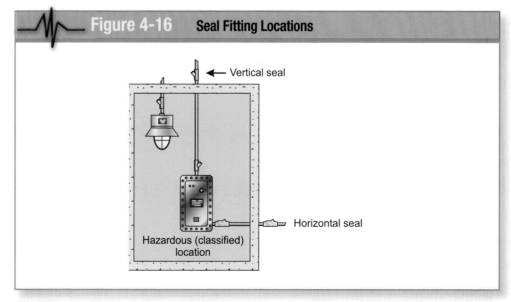

Figure 4-16. *Conduit seals are required at vertical and horizontal boundaries to minimize the passage of gases or vapors through conduit systems to unclassified locations.*

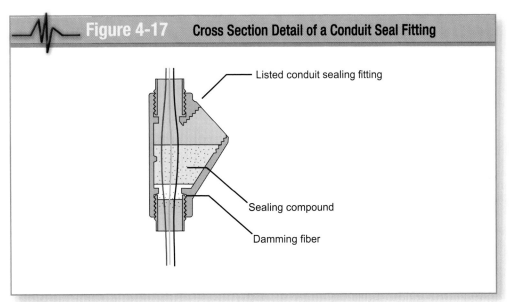

Figure 4-17   **Cross Section Detail of a Conduit Seal Fitting**

- Listed conduit sealing fitting
- Sealing compound
- Damming fiber

Figure 4-17. The damming fiber is installed and the sealing compound is poured to form a tight seal around the individual conductors passing through the fitting.

Class I, Divisions 1 and 2 locations and at the boundaries of hazardous locations. These seals are generally required to be explosionproof, except as permitted by 501.15(B)(2) or 504.70. The fittings are required to be listed for the location. Enclosures can include an integral means for sealing as well. Sealing fittings are required to use a damming material and compound specified by the seal manufacturer. The compound must not be affected by the surrounding atmosphere or liquids, and it is required to have a melting point of not less than 200°F (93°C). The reason for the minimum melting point temperature is to ensure that the seal will be maintained during normal operation.

The listing requirement for these ensures that the sealing fitting, in combination with the appropriate damming material and compound, has been evaluated and listed for providing a seal against the passage of gases or vapors. The seal (with all its components) must have the ability to contain an explosion within an explosionproof enclosure. Sealing fittings are required to be installed in accessible locations and are not permitted to contain any splices or taps.

The thickness of the sealing compound cannot be any less than 5/8 inches (16 mm) and must not be less than the trade size of the fitting. Where conduit sealing fittings are installed, the conduit fill is affected. The conduit fill is limited to not more than 25% of the cross-sectional area of the conduit of the same size, unless the seal fitting is specifically identified for a higher percentage of fill. The reason for the fill limitation is so that an effective sealing can be accomplished by separating the contained conductors during the damming process. **See Figure 4-17.**

Explosionproof Equipment. Equipment enclosed in a case that is capable of withstanding an explosion of a specified gas or vapor that may occur within it and of preventing the ignition of a specified gas or vapor surround the enclosure by sparks, flashes, or explosion of the gas or vapor within, and that operates at such an external temperature that a surrounding flammable atmosphere will not be ignited thereby. [500.2]

Fire Alarm Systems

A fire alarm system is designed to detect hazards associated with fires and notify the building occupants. Fire alarm systems use initiating devices (devices that provide input to the system), a controller (known as a fire alarm control panel or FACP), and notification appliances (output devices). The FACP monitors the initiating devices and displays system status. If any of the initiating devices are triggered the FACP goes into alarm, notifying the occupants with both audible and visual signals.

System Overview

Fire Alarm Control Panel. A fire alarm control panel (FACP) is an electrical panel that is connected to protection system circuits, notification circuits, and other building systems. The FACP is the brain of the system, monitoring circuits for signals and faults. The FACP provides the user interface to indicate system status or system trouble. In the event of a fire the FACP displays the fire's location.

The FACP processes three signal types: alarm signals, supervisory signals, and trouble signals. An alarm signal indicates a fire or other hazardous condition. Supervisory signals indicate an abnormal condition in the fire suppression system. Trouble signals indicate that there is a device, wiring, or power problem. During any of these signal types, the FACP responds accordingly by going into alarm or by providing information to an annunciation panel.

System Components

Initiating Devices and Initiating Device Circuits. An initiating device is a fire protection device that signals a change-of-state condition. An initiating device circuit (IDC) is an electrical circuit consisting of fire alarm initiating devices. Any device in this circuit can activate an alarm signal. Several initiating devices may be connected to a circuit and several circuits may be connected to the FACP. Different circuits may appear as different zones, or areas of

the building. Initiating devices may be the conventional type (non-addressable) using contact closures or the addressable type that can communicate with other addressable devices. Each addressable device in the circuit has its own unique address. This is possible because each addressable device contains its own microprocessor, memory, and software.

Notification Appliances and Notification Appliance Circuits. A notification appliance is a component that provides an audible and/or visual indication of an alarm. A notification appliance circuit (NAC) is an electrical circuit of fire alarm notification appliances that is activated by the FACP. Notification appliances are connected together into a circuit so that they can all be activated simultaneously. Several notification appliance circuits may be connected to the FACP, and the FACP may all be activated at once or by zones. All the notification appliances in a circuit are either all on or all off.

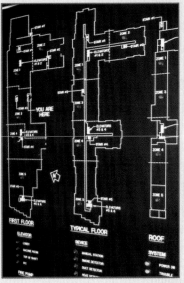

Annunciators. An annunciator panel is used to display fire alarm system information. An annunciator displays fire protection modes, status, trouble, and alarms. The annunciator also displays the fire's location within the building. The annunciator display may be simply lights or LEDs used to represent information, or there may be an alphanumeric (LCD) display. More advanced fire alarm systems may display on computer monitors.

Alarm initiating devices consist of heat detectors, smoke detectors, manual pull stations, and waterflow switches. Notification appliances consist of horns, bells, speakers, strobes or a combination of strobes and speakers on a single device.

Wiring Requirements

Installations of Fire Alarm Systems are usually installed per *NFPA 72, National Fire Alarm and Signaling Code.*

Summary

Although flammable anesthetizing agents and disinfectants are not commonly used in health care facilities currently, the *NEC* continues to include rules that apply to those facilities that allow their use. Where health care facilities use nonflammable anesthetizing agents, fire and explosion hazards and risks are significantly reduced. Where there are no flammable anesthetizing agents, flammable disinfecting agents, or flammable laboratory gases used, there is no requirement for hazardous location classification. These locations are therefore unclassified and general wiring methods and equipment are permitted to be used.

Review Questions

1. __?__ is any area of a facility that has been designated to be used for the administration of any flammable or nonflammable inhalation anesthetic agent in the course of examination or treatment, including the use of such agents for relative analgesia.
   a. A hazardous (classified) location
   b. An unclassified location
   c. An anesthetizing location
   d. A critical care location

2. Which of the following anesthetizing agents is not a flammable type?
   a. Etherethyl chloride
   b. Cyclopropane
   c. Ethyl ether
   d. Halothane

3. Where a room is used for storage of flammable anesthetizing and disinfecting agents, the room is required to be classified in accordance with which of the following?
   a. It is considered an unclassified location
   b. Class I, Division 1 location
   c. Class II, Division 1 location
   d. Class 1, Division 2 location

4. Except as permitted in Section 517.160, all circuits installed in any hazardous flammable anesthetizing location are required to be supplied from which of the following systems?
   a. A corner grounded system
   b. A solidly grounded 3-phase, wye-connected system
   c. A high-impedance grounded neutral system
   d. An isolated power system

5. In an operating room where flammable anesthetics are employed, the entire area shall be considered to be a Class I, Division 1 location that extends upward to a level of __?__ above the floor.
   a. 5 feet (1.52 m)
   b. 3 feet (900 mm)
   c. 4 feet (1.4 m)
   d. The entire room is a Class I, Division 1 location.

6. In an operating room where flammable anesthetics are employed, the area above the hazardous (classified) location is considered to be which of the flowing?
   a. A Class I, Division 2 location
   b. A Class I, Division 1 location
   c. An area above a hazardous (classified) location
   d. A Class I, Zone 0 location

7. Receptacles and attachment plugs in a hazardous (classified) location(s) employing flammable anesthetizing agents shall be listed for use in __?__ and shall have provision for the connection of a grounding conductor.

   a. Class I, Group C hazardous (classified) locations

   b. Class I, Group D hazardous (classified) locations

   c. Class II, Group E hazardous (classified) locations

   d. Class III locations

8. Which of the following wiring methods are not permitted for use above a hazardous (classified) location referred to in 517.60?

   a. Rigid metal conduit

   b. Electrical metallic tubing

   c. Intermediate metal conduit

   d. Nonmetallic sheathed cable

9. Generally, which of the following wiring methods is permitted to be installed other-than-hazardous (classified) locations, as provided in 517.60?

   a. Any metal raceway system or cable assembly

   b. A metal raceway system or cable assembly that qualifies as an equipment grounding conductor path in accordance with 250.118

   c. A nonmetallic wiring method that includes an equipment grounding conductor sized in accordance with 250.118

   d. None of the above wiring methods are permitted

10. In an anesthetizing location, which of the following is not required to be grounded?

    a. Metal raceways

    b. Metal cable assemblies

    c. Fixed electrical equipment

    d. Equipment operating at not more than 10 volts

11. Low-voltage equipment and instruments that are frequently in contact with the bodies of persons or has exposed current-carrying elements shall comply with which of the following?

    a. They operate on an electrical potential of 10 volts or less

    b. They are approved as intrinsically safe or double-insulated equipment

    c. They are moisture resistant

    d. Any of the above

12. All 120-volt, single-phase, 3-wire, 2-pole grounding type receptacles and attachment plugs located above hazardous (classified) anesthetizing locations are required to meet which of the following requirements?

    a. The receptacles and attachment plugs shall be listed for hospital use

    b. They are required to be listed isolated grounding types

    c. They are required to be listed hospital grade receptacles and attachment plugs

    d. Any standard grounding-type receptacle is permitted in these locations

13. What is the minimum thickness of the compound installed within a conduit seal fitting?

    a. $^3/_8$ in.

    b. $^1/_2$ in.

    c. $^5/_8$ in.

    d. $^3/_4$ in.

14. According to Table 1 of Chapter 9, a typical conduit installation allows a maximum conduit fill of 40% for two or more conductors. Unless the conduit seal is specifically identified otherwise, what is the maximum conduit fill for two or more conductors where a conduit seal fitting has been installed in a run of conduit?

    a. 25%

    b. 30%

    c. 35%

    d. None of the above

15. How can the installation of electrical equipment suitable for hazardous locations including sealing fittings be avoided for anesthetizing locations?

    a. Where there are no flammable anesthetics used

    b. Where there are no flammable disinfectants used

    c. Where there are only non-flammable gases used

    d. All of the above

Diagnostic Imaging Equipment Installations

The *NEC* refers only to X-ray installations, but modern health care facilities include other tools for taking pictures of the human body. The latest medical term for this class of equipment is diagnostic imaging. Magnetic resonance imaging (MRI) uses magnetic fields, and ultrasound imaging uses sound waves to generate pictures without X-rays. Even the machines that still use X-rays have evolved into much more complex instruments than they were before. The CT scan, for instance, uses computer tomography to generate a 3-D image from multiple 2-D X-ray images. In addition, there are also treatment machines that use diagnostic imaging for precision body cell alignment along with gamma rays and even alpha particles to treat these precision aligned cells. The *NEC* refers to these types of machines as therapeutic X-ray equipment. Both diagnostic and therapeutic X-ray equipment are covered in Part V of Article 517. From the perspective of how these machines are connected to (or disconnected from) the health care facility's electrical system, however, the rules of the *Code* still apply.

Objectives

» Distinguish the requirements for connecting fixed and portable imaging equipment to the supply circuit

» Summarize the requirements for disconnecting imaging equipment from the supply circuit

» Define the requirements for overcurrent protection provided at branch and feeder circuits for diagnostic imaging equipment

» Explain circuit conductor requirements for supplying imaging equipment

» Determine guarding and grounding requirements for imaging equipment

Chapter 5

Table of Contents

INTRODUCTION

Wiring techniques and requirements for X-ray equipment, magnetic resonance imaging equipment, and computerized axial tomography equipment are covered in several places of the *Code*. First, all the general practices of good electrical installations covered in Chapters 1-4 of the *Code* apply, except as modified by Article 517. Next, all the general wiring considerations applicable to health care facilities covered in Part II of Article 517 apply. These have already been discussed in earlier chapters of this book.

It is important to notice another article within the *NEC* that covers X-ray equipment, Article 660 X-Ray Equipment. However, Article 660 does not cover X-ray equipment related to medical procedures. Rather it covers industrial X-ray devices, so this article is of no significance to this book or to the study of X-ray equipment in health care facilities.

Finally, Article 517, Part V X-Ray Installations covers the installation of X-ray equipment in health care facilities. Part V of Article 517 is the major focus of this chapter.

Radiation therapy has been in use as a cancer treatment for more than 100 years, with its earliest roots traced from the discovery of X-rays in 1895 by Wilhelm Röntgen.

The field of radiation diagnostics and therapy began to grow in the early 1900s, largely due to the groundbreaking work

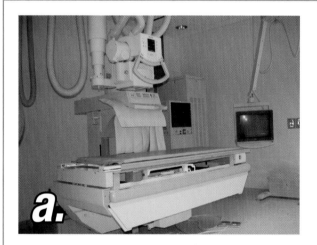

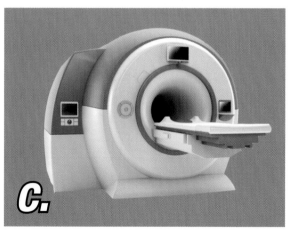
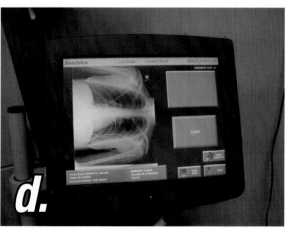

Simple X-ray equipment (a) generates 2-D images while CT scanning equipment (b) uses X-rays and computer tomography to generate 3-D images. MRI equipment (c) uses magnetism instead of X-rays. Digital technology makes the results of body scans and X-rays viewable almost immediately (d). Technology evolves; so does the Code.

Safe use of X-ray and other imaging technologies (other than the electrical wiring, connections and installations of disconnect switches, and related overcurrent protection and associated control circuits) is beyond the scope of this book, as well as beyond the scope of the *NEC*. The *Code* specifically states, at the beginning of Part V, Article 517, "Nothing in this part shall be construed as specifying safe-guards against the useful beam or stray X-ray radiation." Further, Informational Note No. 1 explains that radiation safety and performance requirements of several classes of X-ray equipment are regulated under Public Law 90-602 and are enforced by the Department of Health and Human Services. [Article 517, Part V, Informational Note No. 1]

of Nobel Prize-winning scientist Marie Curie, who discovered the radioactive elements polonium and radium. This began a new era in medical treatment and research. Radium was used in various forms until the mid-1900s when cobalt and caesium units came into use. Medical linear accelerators have been used, too, as sources of radiation since the late 1940s.

With Godfrey Hounsfield's invention of computed tomography (CT) in 1971, three-dimensional planning became a possibility and created a shift from 2-D to 3-D radiation delivery. CT-based planning allows physicians to more accurately determine the dose distribution using axial tomographic images of the patient's anatomy. Orthovoltage and cobalt units

have largely been replaced by megavoltage linear accelerators, useful for their penetrating energies and lack of physical radiation source.

The advent of new imaging technologies, including magnetic resonance imaging (MRI) in the 1970s and positron emission tomography (PET) in the 1980s has moved radiation therapy from 3-D conformal to intensity-modulated radiation therapy (IMRT) and image-guided radiation therapy (IGRT). These advances allowed radiation oncologists to better see and target tumors, which has resulted in better treatment outcomes, more organ preservation, and fewer side effects.

Though the *Code* only specifically refers to X-ray installations, the wiring requirements apply to a large family of medical imaging equipment in use today and are therefore considered to apply unless otherwise stated.

As new inventions enter the health care field, the requirements for connecting them to the electrical supply evolve. The *Code* specifically allows for manufacturer directions to supplement its requirements, thus allowing different kinds of equipment to be installed according to *Code* without requiring the *Code* itself to anticipate all the technologies that may come to exist. In effect, the manufacturer's instructions become the rule.

CONNECTION TO SUPPLY CIRCUIT

Connection requirements depend on whether the equipment being connected is fixed or portable and on whether the supply circuit is less than 600 volts, or over 600 volts.

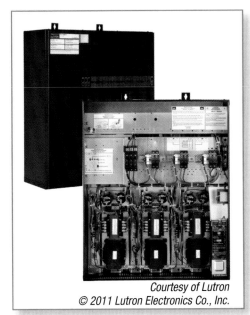

Courtesy of Lutron
© 2011 Lutron Electronics Co., Inc.

DCI dimming panels meet recommended levels of noise of MRI system manufacturers.

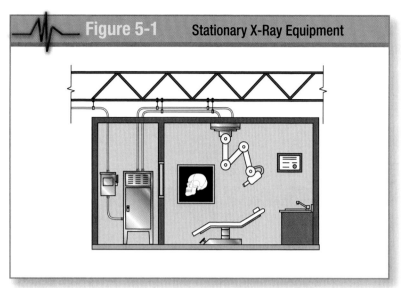

Figure 5-1. *X-ray equipment generally is permitted to be wired using any of the wiring methods of Chapters 1 through 4 of the NEC.*

Fixed and Stationary Equipment

Fixed and stationary equipment must be connected to the power supply by means of a wiring method that complies with the requirements of Chapters 1-4 of the *Code*, as modified by Article 517. **See Figure 5-1.**

However, for portable, mobile, or transportable equipment properly supplied by a branch circuit rated at not over 30 amperes, medical X-ray equipment may be supplied through a suitable attachment plug and hard-service cable or cord. **See Figure 5-2.**

Portable, Mobile, and Transportable Equipment

Portable, mobile, and transportable medical imaging equipment is widely used in health care facilities. Though the three terms may be used loosely to mean the same thing, they do have technical distinctions as follows:

- *Portable X-ray installations* are designed to be hand carried.
- *Mobile X-ray installations* are mounted on a permanent base with wheels, casters, or a combination of both to facilitate moving the equipment while completely assembled.
- *Transportable X-ray installations* are designed to be conveyed by a vehicle or readily disassembled for transport by a vehicle.

Portable, mobile, and transportable medical X-ray equipment requiring a capacity of not over 60 amperes is not required to be supplied by an individual branch circuit. It is allowed, but not required. **See Figure 5-3.**

Over 600 Volt Supply

The general requirements for equipment operated on a supply circuit of over 600 volts must comply with Article 490, "Equipment, Over 600 Volts, Nominal." 517.71(C) provides a reference to Article 490 only, but the requirements for

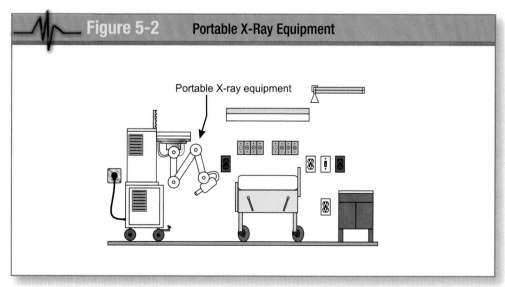

Figure 5-2. *Portable X-ray equipment is permitted to be cord- and plug-connected.*

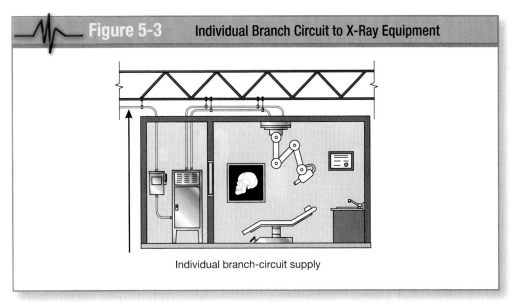

| Figure 5-3 | Individual Branch Circuit to X-Ray Equipment |

Individual branch-circuit supply

Figure 5-3. *Individual branch circuits serve only one piece of equipment. They are permitted, but not required, for equipment operating under 60 amperes.*

circuits at these voltages are found in other articles and these circuit installations are required to meet all applicable rules. Article 310, for instance, provides specific information and rules relating to conductors rated over 600 volts. General requirements for wiring methods used with systems over 600 volts are provided in Part II of Article 300. Equipment that operates at over 600 volts is also covered in Part II of Article 110 which includes working space requirements for this type of equipment. However, this study of health care facilities is limited to 600 volts, nominal or less.

DISCONNECTING MEANS

Disconnecting means that are required by the Code for any equipment is typically related to worker safety, and the required disconnect for X-ray equipment is no exception. X-ray equipment is required to have a disconnecting means. **See Figure 5-4.**

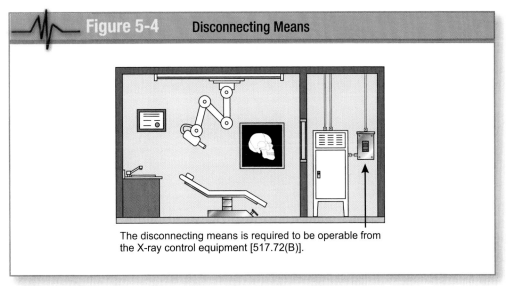

| Figure 5-4 | Disconnecting Means |

The disconnecting means is required to be operable from the X-ray control equipment [517.72(B)].

For additional information, visit qr.njatcdb.org Item #1055

Figure 5-4. *A disconnecting means is required for X-ray equipment.*

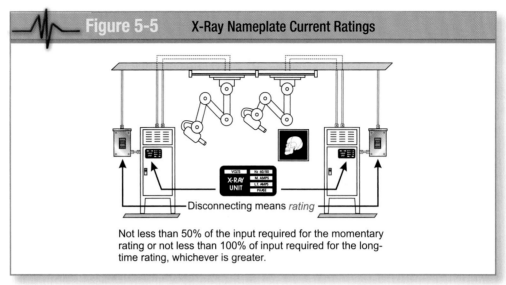

Figure 5-5. Nameplates of X-ray equipment provide momentary and long-time current ratings used for sizing disconnecting means.

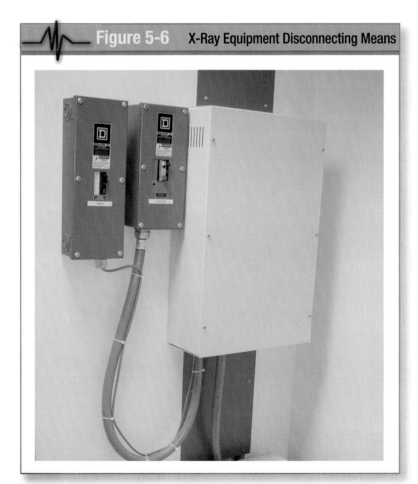

Figure 5-6. An enclosed circuit breaker is often installed as the disconnecting means for X-ray equipment.

This disconnecting means provides service personnel a means by which to place the equipment in an electrically safe work condition while service operations and maintenance procedures are performed. It also serves as a ready means by which an operator may disconnect the equipment in case of a failure or abnormal condition related to the equipment's use.

Capacity

As for any equipment covered by the *Code*, the required disconnecting means installed in the supply circuit for X-ray equipment must be sufficient to safely interrupt the circuit it is intended to disconnect. This circuit is going to be subject to two levels of current in normal operation. The two ratings that must be addressed are the equipment's momentary rating and its long-time rating. The *momentary rating* is a rating based on an operating interval that does not exceed 5 seconds. This can be compared to an inrush starting current of an electric motor, and it is typically a high level of current for a short period of time. The *long-time rating* refers to the ongoing (periods of 5 minutes or longer) operating levels of current required after startup. The disconnecting means must have a rating not less than 50% of the input required for the momentary rating or not less than 100% of input required for the long-time rating of the equipment, whichever is greater. This applies to each disconnecting means in the supply circuit.

The equipment should have a data nameplate that provides the equipment manufacturer's specific ratings for the installations. **See Figure 5-5.**

It is important to refer to the requirements provided by the equipment manufacturer because, especially with medical equipment, these may exceed the *Code* minimums.

The disconnecting means can be a fused disconnect or an enclosed circuit breaker. X-ray equipment manufacturers generally specify an enclosed circuit breaker. **See Figure 5-6.**

Location

The disconnecting means must be operable from a location readily accessible from the X-ray control. [517.72(B)] *Readily accessible* is defined as capable of being reached quickly for operation, renewal, or inspections without requiring those to whom ready access is requisite to climb over or remove obstacles or to resort to portable ladders, and so forth. This means the equipment's operator must be able to access the supply circuit disconnect quickly for operation without having to

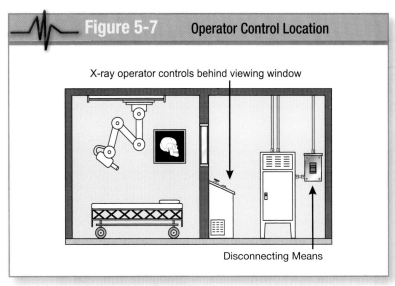

Figure 5-7. The X-ray equipment disconnecting means is required to be readily accessible by the operator from the control location.

climb over or remove obstacles or to resort to portable ladders from the location where the X-ray unit is controlled during patient treatment. **See Figure 5-7.**

There are no exceptions to this disconnect location requirement, unlike other sections of the Code which sometimes permit the disconnecting means for other kinds of equipment to be out of sight. **See Figure 5-8.**

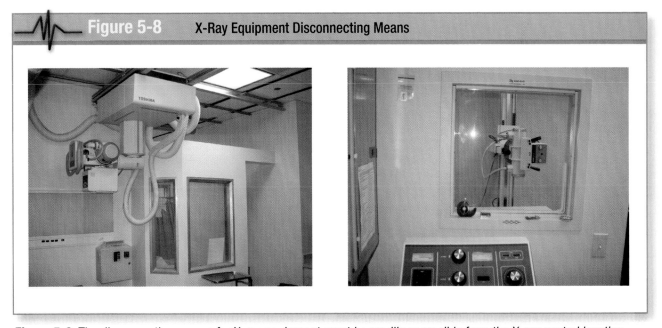

Figure 5-8. The disconnecting means for X-ray equipment must be readily accessible from the X-ray control location.

Figure 5-9   X-Ray Control Console

Figure 5-9. *As positioned, the X-ray control console creates an obstruction of the required working space for the panelboard, but the console is not fixed in its position and can be moved away from the panelboard if necessary.*

Not only must the disconnection means be accessible to the equipment operator, it must also be accessible to the electrical worker who must work on it. This means equipment controls must either be positioned far enough from the disconnection means to allow work space, or the equipment must be on wheels capable of being rolled easily out of the way. **See Figure 5-9.**

Portable Equipment

More and more facilities are using mobile or portable medical equipment.

> **X-Ray Installations, Portable.** X-ray equipment designed to be hand carried.
>
> **X-Ray Installations, Transportable.** X-ray equipment to be conveyed by a vehicle or that is readily disassembled for transport by a vehicle.

Health care facilities still have designated areas for radiology treatment and other imaging technologies that require specialized constructions. However, as imaging technology has advanced, X-ray equipment has evolved from mostly fixed and stationary types to portable units that can be taken to the patient. Portable units require special supply outlets and disconnecting means. Where portable X-ray equipment is connected to a 120-volt branch circuit rated at 30 amperes or less, the disconnecting means can be the same attachment plug and receptacle used to supply the equipment with power. Portable equipment possesses a cord and attachment plug as an integral part of its unit. Health care facilities where this type of X-ray equipment is utilized are usually designed with appropriate circuits and receptacles at specified locations. These receptacles and attachment plugs must be evaluated and listed for this use. **See Figure 5-10.**

RATING OF SUPPLY CONDUCTORS AND OVERCURRENT PROTECTION

The general rule given by the *Code* in Article 110.3(B) for installing electrical equipment is that listed or labeled equipment must be installed and used in accordance with any instructions included in the listing or labeling. This allows manufacturers to refine the details without changing the *Code*. That said, however, the *Code* does provide some guidelines for supply conductors and overcurrent protection which vary depending on whether the equipment is used for diagnostic purposes or therapeutic purposes.

Diagnostic Equipment

As mentioned earlier, equipment continues to evolve, and so do the requirements for connecting it to the electrical supply. The manufacturer's instructions generally provide very specific rules, but in the absence of any other instructions, the *Code* provides the following general requirements for diagnostic equipment:

- *Branch Circuits.* The ampacity of supply branch-circuit conductors and the current rating of overcurrent protective devices must not be less than 50% of the momentary rating or 100% of the long-time rating, whichever is greater.
- *Feeders.* Where a feeder supplies two or more branch circuits, the feeder must have adequate capacity for the anticipated mode of operations. The ampacity of supply feeders and the current rating of overcurrent protective devices supplying two or more branch circuits supplying X-ray units must not be less than 50% of the momentary demand rating of the largest unit plus 25% of the momentary demand rating of the next largest unit plus 10% of the momentary demand rating for each additional unit. Where simultaneous biplane examinations are undertaken with the X-ray units, the supply conductors and overcurrent protective devices must be 100% of the momentary demand rating of each X-ray unit.

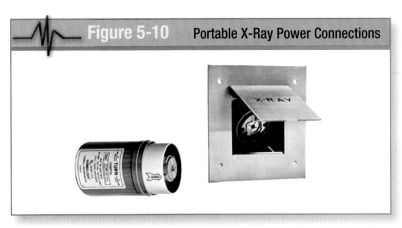

Figure 5-10. Portable X-ray equipment uses specialized attachment plugs and receptacles.

Where an X-ray equipment feeder originates in a service switchboard, and that service is equipped with ground-fault protection in accordance with Section 230.95 or 215.10, the X-ray equipment feeder is considered as a second level circuit breaker and also is required to be protected by a ground-fault protection in accordance with 517.17(B). **See Figure 5-11.**

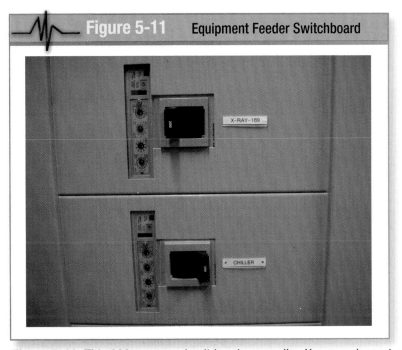

Figure 5-11. This 200-ampere circuit breaker supplies X-ray equipment in a hospital. This feeder breaker is also equipped with ground-fault protection features in compliance with NEC 517.17(B).

The minimum conductor size for branch and feeder circuits is also governed by voltage regulation requirements. For a specific installation, the manufacturer usually specifies minimum distribution transformer and conductor sizes, rating of disconnecting means, and overcurrent protection. Minimum voltages may also be specified within the manufacturer's site guide and installation instruction.

Therapeutic Equipment

With the ongoing invention of new therapeutic equipment, the requirements for connecting such equipment to the electrical supply also evolve. The general rule is that the ampacity of conductors and rating of overcurrent protective devices must not be less than 100% of the current rating of the equipment. The ampacity of the branch-circuit conductors and the ratings of disconnecting means and overcurrent protection for the equipment are usually designated by the manufacturer for the specific installation. The *Code* indicates that the manufacturer's instructions become the rule in these cases.

CONTROL CIRCUIT CONDUCTORS

X-ray equipment often requires significant control circuit wiring and interconnection. Control circuit conductors may need to meet raceway fill and specific size requirements. The fundamental requirements regarding raceway fill include Section 300.17, which states that the number and size of control circuit conductors in any raceway must not be more than will allow heat to dissipate.

Ready installation or withdrawal of the conductors must be possible without damage to them or their insulation. In addition, control circuit conductors may need to meet the many requirements of Article 725 depending on specific applications or installation techniques.

Also, size 18 AWG or 16 AWG fixture wires and flexible cords are allowed for the control and operating circuits of X-ray and auxiliary equipment where protected by overcurrent devices no larger than 20 amperes. [517.74(B)]

EQUIPMENT INSTALLATION APPROVALS

All equipment for new installations and all used or reconditioned equipment moved to and reinstalled at a new location must be of an approved type. As defined in Article 100, the word *approved* means "acceptable to the authority having jurisdiction." When *Code* users encounter this situation, it is always a good practice to communicate with the approving authority about the specific criteria of such installations. [517.75] As a general rule, authorities having jurisdiction base their approval of equipment on whether it is listed or certified by a qualified electrical testing laboratory. There are a number of qualified electrical testing laboratories that also provide field evaluation services to assist owners, design teams, and installers where approvals are required. Listing or field evaluation by nationally recognized testing laboratories does not constitute an approval, but it does provide a strong basis for the inspector to issue such approvals.

TRANSFORMERS AND CAPACITORS

Any transformer or capacitor installed as an integral part of the X-ray equipment is not required to comply with the rules in Articles 450 and 460. This is a good example of a rule in Article 517 that modifies a general provision in Chapter 4 of the *Code*. Capacitors that are part of X-ray equipment must be installed in metal enclosures that are grounded through a connection to the equipment grounding conductor of the supply circuit or be mounted in enclosures that are made of insulating material. These transformers and capacitors that are integral components of the X-ray equipment are usually provided by the manufacturer with specific direction on installation and wiring.

INSTALLATION OF HIGH-TENSION X-RAY CABLES

High-tension X-ray equipment cabling with grounded shields that connect X-ray tubes and image intensifiers are permitted to be installed in wireways (troughs) or cable trays along with the X-ray equipment control and power supply conductors. Where so installed, separation is not required, unless the manufacturer of the equipment specifically requires separation. Where separation of conductors is required, wireways or cable trays with suitable fixed barriers can be installed to meet the requirement.

GUARDING AND GROUNDING

X-ray equipment requires careful grounding to protect users, workers, and patients from electrical shock. In addition to the *NEC* grounding and bonding requirements discussed in the later section Non-Current-Carrying Metal Parts, the site planning booklet and the manufacturer's instructions will require careful review for possible additional or increased grounding and bonding.

High-Voltage Parts

All high-voltage parts, including X-ray tubes, must be mounted within grounded enclosures. Air, oil, gas, or other suitable insulating media must be used to insulate the high-voltage from the grounded enclosure. The connection from the high-voltage equipment to X-ray tubes and other high-voltage components must be made with high-voltage shielded cables. These cables must be rated in accordance to the voltage of the equipment. The manufacturer's data nameplates and instructions typically provide information about the

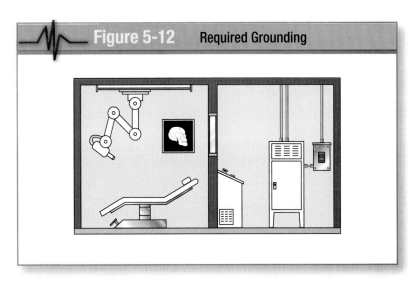

Figure 5-12. *All non–current-carrying metal parts of X-ray and associated equipment are required to be grounded by connecting to the equipment grounding conductor of the X-ray equipment supply circuit.*

minimum voltage ratings for such cables or conductors.

Low-Voltage Cables

Low-voltage cables connecting to oil-filled units that are not completely sealed, such as transformers, condensers, oil coolers, and high-voltage switches, must have insulation of the oil-resistant type.

Non-Current-Carrying Metal Parts

Since patients interact with X-ray equipment, the equipment is considered to be part of the patient care area. Therefore, all non-current-carrying metal parts of X-ray and associated equipment (controls, tables, X-ray tube heads, etc.) must be connected to an equipment grounding conductor in the manner specified in Article 250, as modified by 517.13(A) and (B) for redundant equipment grounding purposes. **See Figure 5-12.**

Local Area Networks and Structured Cabling Systems

The use of electronic medical records is an essential tool in hospitals, medical practices, and other health care facilities. Access to these electronic records allows for a reliable method of communication between physicians, nurses, labs, and other clinical staff without relying on handwritten notes or files stored as paper records in single-location filing room. Patient information can be accessed from multiple computers from locations with password-protected security. To maintain electronic records and the high security required by state and federal authorities requires robust file servers, routers, switches, and a reliable communication network or infrastructure. The system that makes up this network is known as a local area network or LAN. When multiple buildings are connected together (as in a hospital or college or university) to the same network, this network is known as a campus area network. When metropolitan areas, regional areas, countries, etc. are connected to a network, this is known as a wide area network or WAN.

System Overview

Networks used to process data, such as medical records, start with a reliable passive cabling system which connects active components such as computers, servers, routers, switches, and perhaps other building systems such as components of a building automation system. Over the years this cabling system has become more or less standardized and is known as a structured cabling system (SCS). Structured cabling systems consist of unshielded (UTP) or shielded (FTP) twisted copper cabling, and multimode and or single mode fiber optic cable. Coaxial cable may also be part of a structured cabling system.

A structured cabling system is divided into seven standard elements. These elements are defined in the ANSI/TIA/EIS *Commercial Building Telecommunications Cabling Standard* 568-B.1.5. These elements are 1) horizontal cabling, 2) backbone cabling, 3) work area, 4) telecommunications rooms, 5) equipment rooms, 6) entrance facilities, and 7) administration.

System Components

We will describe a structured cabling system starting with the work area, and then working our way back to the service entrance where the service providers supply their signal and services to the building.

Work Area. The work area consists of the cabling (cords and/or adapters), which extends from the telecommunications outlet to the station equipment. Station equipment may include any number of devices including, but not limited to, telephone sets, data terminals, personal computers, workstations, video monitors, video cameras, and facsimile machines. In other words, the work area is where the computer connects to the cabling system and then to the other active devices on the LAN.

Horizontal Cabling. Horizontal cabling includes the outlets at the work areas, the cables from the work areas to the telecommunications rooms, the termination of the cables in the telecommunications rooms, and horizontal cross-connections within the telecommunications rooms. The term *horizontal* is used because these cables usually run horizontally in floors or ceilings and do not penetrate a floor.

Telecommunications Room. A Telecommunications Room (TR) is an area set aside to house equipment associated with a telecommunications cabling system. The primary function of a telecommunications room is the termination of horizontal and backbone cables to compatible connecting hardware.

Backbone Cabling. The purpose for backbone cabling is to provide interconnections between telecommunications rooms, equipment rooms, and entrance facilities. Intra-building backbone cables are the cables that exist within a building. Cables between buildings are called interbuilding backbone cables. Backbone cabling consists of backbone cables, intermediate and main cross-connects, mechanical terminations, and patch cords or jumpers for backbone-to-backbone cross-connections. Backbone cables include both fiber and unshielded twisted pair cables.

Equipment Room. An equipment room is an area where the large telecommunications systems equipment that is shared by many users is housed. Equipment typically found in an equipment room includes voice switches, data switches, front-end processors terminal controllers, LAN servers, LAN switches, and LAN routers. Equipment rooms are distinct from telecommunications rooms because of the nature of the equipment (which is usually very large and serves all users in a building) they contain. However, the function of a telecommunications room may also be performed in an equipment room.

Entrance Facility. An entrance facility consists of cables, connecting hardware, protection devices, and other equipment needed to connect cables entering from outdoors to cables which are suitable and approved for use indoors. An entrance facility may be in a room by itself, or it may be collocated with an equipment room or with a telecommunications room.

Administration. Administration includes documentation (labels, records, drawings, reports, and work orders) of cables, termination hardware, patching and cross-connect facilities, pathways, telecommunications rooms, and spaces. Cross-connections and interconnections are performed at administration points in the backbone and horizontal cabling systems. Cross-connections are made with twisted pair hook-up wire or patch cords. A hook-up wire is a short length of unjacketed solid conductor twisted pair(s).

Wiring Requirements

Wiring used in standards based structured cabling systems include unshielded twisted pair (UTP), shielded twisted pair (STP, ScTP, FTP or PiMF) and fiber optic cable, both multimode and single mode.

An unshielded twisted pair (UTP) is a pair of solid copper conductors that are twisted together in a manner that minimizes crosstalk (interference) with other pairs of wires in the same unshielded cable. A UTP cable consists of four pairs of copper conductors.

The use of shielded cable is reemerging as a viable option because of the demand for twisted-pair cables to perform at higher frequencies; this will allow for greater bandwidth. New naming conventions are being used to better describe the types of shielded cable available.

ScTP. Screened Twisted Pair utilizes a metallic screen or braid over the group of four pairs

FTP. Foil Twisted Pair is similar to ScTP but uses a thin metallic foil as opposed to a braid.

PiMF. Pairs in Metal Foil cables have a foil shield for each pair to virtually eliminate internal crosstalk and usually include an overall foil or screen to provide additional EMI immunity.

Multi-Mode Fiber (MMF) is an optical fiber that can carry multiple modes of light at the same time down the fiber. Although multi-mode fiber cabling has typically been installed in premises networks, more single-mode fiber is being installed for the same purpose.

Single-Mode Fiber (SMF) is an optical fiber where a signal in the form of a single mode of light travels along one path in the fiber.

Summary

Special rules apply to electrical wiring for X-ray equipment for health care use. Article 660 of the *NEC* covers all X-ray equipment operating at any frequency or voltage for industrial or other non-medical or non-dental uses. Part V of Article 517 covers installations of X-ray equipment and associated electrical wiring installations in health care facilities, including dental treatment centers. Part V of Article 517 includes rules that guide installers and engineering teams on the required sizes for conductors, locations of equipment disconnects, and overcurrent protection sizes for X-ray equipment. X-ray equipment can vary in size from very large fixed units to smaller, more portable equipment. These days in hospitals and other health care facilities, both fixed and portable types of X-ray equipment are being used. It is not uncommon to see X-ray equipment brought to a patient care room, rather than bringing the patient to the X-ray equipment. Wiring requirements for patient care rooms have been covered in earlier chapters of this book, but in cases where X-ray equipment may be used in those rooms, the requirements of this chapter must also be considered when wiring those rooms.

Safety practices related to using X-ray equipment and technology is beyond the scope of this book. Radiation safety and performance requirements for several classes of X-ray equipment are regulated under Public Law 90-602 and are enforced by the Department of Health and Human Services. [Article 517, Part V, Informational Notes No. 1 and No. 2]

Review Questions

1. **Which of the following is not covered by the requirements of Part V of Article 517 that apply to X-ray equipment?**
   a. Disconnection means locations
   b. Sizing conductors for X-ray equipment
   c. Overcurrent protection for X-ray equipment
   d. Radiation safety and performance requirements of several classes of X-ray equipment

2. **The long-time rating of X-ray equipment is based on an operating interval of ? .**
   a. 5 minutes or longer
   b. 5 seconds or longer
   c. 1 hour
   d. continuous duty

3. **The momentary rating of an X-ray equipment installation is a rating based on an operating interval that does not ? .**
   a. exceed 60 seconds
   b. exceed 10 seconds
   c. exceed 5 minutes
   d. exceed 5 seconds

4. **Which of the following wiring methods are generally permitted to be used for fixed and stationary X-ray equipment installations in health care facilities?**
   a. Rigid metal conduit or intermediate metal conduit
   b. Rigid nonmetallic conduit
   c. Electrical metallic tubing
   d. All of the above

5. **Individual branch circuits shall not be required for portable, mobile, or transportable medical X-ray equipment requiring a capacity that is not in excess of ? .**
   a. 30 amperes
   b. 50 amperes
   c. 60 amperes
   d. 100 amperes

6. The disconnection means for X-ray equipment is required to have a rating at least __?__ of the input required for the momentary rating or __?__ of the input required for the long-time rating of the equipment, whichever is greater.

   a. 50% / 100%

   b. 100% / 50%

   c. 125% / 80%

   d. 80% / 125%

7. The disconnection means for X-ray equipment is required to be located readily accessible from which of the following?

   a. The switchboard supplying the equipment

   b. The X-ray table

   c. The X-ray equipment room

   d. The X-ray control

8. Unless otherwise designated by the manufacturer of the X-ray equipment, the ampacity of conductors and rating of overcurrent protective devices for medical X-ray therapy equipment shall not be less than which of the following?

   a. 125% of the current rating

   b. 100% of the current rating

   c. 80% of the current rating

   d. 150% of the current rating

9. The supply branch circuit conductors and the current rating of the overcurrent protective devices for X-ray equipment are required to be at least __?__ of the input required for the momentary rating or __?__ of the input required for the long-time rating of the equipment, whichever is greater.

   a. 50% / 100%

   b. 100% / 50%

   c. 125% / 80%

   d. 80% / 125%

10. All equipment for new X-ray installations and all used or reconditioned X-ray equipment moved to and reinstalled at a new location shall be of a(n) __?__ type.

    a. Identified

    b. Listed

    c. Certified

    d. Approved

11. Where a manufacturer of X-ray equipment requires a minimum size of circuit conductor of 100 amperes, what is the minimum size aluminum XHHW-2 conductors that can be used to supply the equipment? *Note, the terminals marked on the disconnect and the X-ray equipment are identified for 60/75° temperature rating.*

    a. 2 AWG

    b. 1 AWG

    c. 1/0 AWG

    d. 2/0 AWG

12. Where the overcurrent protective device for X-ray equipment is rated at 125 amperes, what is the minimum size copper equipment grounding conductor (wire-type) that is required?

    a. 8 AWG

    b. 6 AWG

    c. 4 AWG

    d. 1 AWG

13. Which of the following *NEC* Sections does not cover medical X-ray equipment?

    a. 110.3(B)

    b. 240.4

    c. 300.17

    d. 660.1

14. Which of the following *Code* sections requires a separate and redundant equipment grounding conductor in the branch circuit of a metal raceway supplying a piece of X-ray equipment?

    a. 110.3(A)

    b. 250.118(5)

    c. 344.60

    d. 517.78(C)

15. Which of the following selections best describes the reason for installing redundant equipment grounding conductors in a supply conduit for an X-ray machine circuit?

    a. Medical X-ray machines are marked as such

    b. Medical X-ray machines are used during patient treatment

    c. Medical X-ray equipment are used most often in radiology

    d. Medical X-ray machines are sometimes used in other areas of a hospital

Isolated Power Systems

Installations of isolated power systems in hospitals date back to the 1940s. At that time they were used primarily to reduce electrical arcing in areas where flammable anesthetics were employed. Though flammable anesthetics are no longer an issue, isolated power systems continue to be an important part of a health care facility's supply system because they offer other advantages. In particular, they provide protection from electrical shock without allowing any interruption in service due to ground faults. This makes them particularly useful in wet procedure areas, such as operating rooms, where interrupting a critical procedure could prove devastating.

Objectives

» Explain the benefits of using an isolated power system

» List the major electrical components of an isolated power system and describe how they are wired

» Describe the equipment grounding conductor connections in the isolated power panel and at the isolated power receptacles

» Describe how line isolation monitors work

Chapter 6

Table of Contents

INTRODUCTION

Isolated power systems are covered by Part VII of Article 517. They offer advantages because they are electrical systems that operate ungrounded. An ungrounded system means no secondary conductor supplied by an isolation transformer is intentionally grounded. An *isolation transformer* is a transformer of the multiple-winding type, with the primary and secondary windings physically separated, which inductively couples its secondary winding(s) to circuit conductors connected to its primary winding(s). Sections 250.22(2) and 517.160(A)(1) and (2) specifically restrict isolated power systems from being grounded. Normally electrical systems are expected to be grounded, and in wet locations especially, they use ground-fault circuit-interrupting devices to protect people from stray electrical currents. However, in certain critical areas of a health care facility, such as an operating room, which are also often considered wet procedure locations, where any interruption of electrical supply could prove fatal to a patient, system grounding and ground-fault circuit interrupters are not allowed.

THE BENEFITS

The principle benefit of operating an ungrounded system in today's health care facility is continuity of service should a ground-fault condition occur. Medical and nursing sciences are becoming progressively more dependent on electrical apparatus for the preservation of life. Every year more cardiac operations are performed, during some of which the patient's life depends on artificially circulating the blood. In other operations, life is sustained by means of electric impulses that stimulate and regulate heart action. At the same time, the equipment, the doctors, and the nurses may all be standing in prepping solution and other conductive fluids that may greatly increase the risk of shock. These locations are defined in 517.2 as wet procedure locations.

In the past, when flammable anesthetics were used, the advantage of using an ungrounded system was also that such a system was unlikely to produce an arc to ignite the flammable atmosphere. Today, flammable anesthetics are no longer a worry, but other reasons make isolated power systems indispensable. They provide enhanced safety for patients and operational staff by:

- Reducing shock hazard
- Maintaining power continuity
- Reducing electrical noise
- Providing advance warning of equipment failure

Isolated power systems monitor current leakage as does a GFCI, but instead of using a GFCI device to interrupt power immediately, they issue a warning signal when leakage current reaches a defined limit. The warning signal allows the procedure to continue and provides time for the attending medical professionals to replace the offending equipment or device in an orderly fashion. This eliminates the shock hazard before it becomes too high, without a GFCI power interruption. To provide the power redundancy needed in some situations, multiple isolated power systems can be installed.

INSTALLATIONS OF ISOLATED POWER CIRCUITS

In critical care locations, where interruption by a GFCI cannot be tolerated, the isolated power system becomes a mandatory *Code* requirement [517.20(A)(1)]. An exception to the requirement in 517.20(A) relaxes the rule for branch circuits that supply listed, fixed single- or 3-phase therapeutic, and diagnostic equipment and are fed from a grounded system, as long as the wiring from the grounded system and isolated power system circuits are not contained in the same raceway or enclosure, and all conductive surfaces of equipment are connected to an insulated equipment grounding conductor. [517.20(A) Exception]

Wherever the design specifies installing an isolated power system, it is required to be listed and must be installed in accordance with the rules contained in 517.160 and in accordance with any instructions included by the manufacturer of the system. Each isolated power circuit must be controlled by a switch that has a

---

Fact

An excellent source of additional information is the technical white paper published by Schneider Electric (Square D) entitled "Isolated Power Systems (IPS) and Wet Locations in Healthcare Facilities."

For additional information, visit qr.njatcdb.org
Item #1086

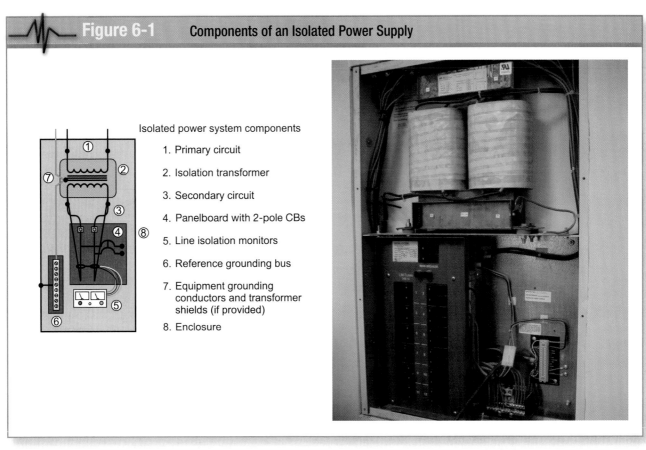

Figure 6-1 **Components of an Isolated Power Supply**

Isolated power system components

1. Primary circuit
2. Isolation transformer
3. Secondary circuit
4. Panelboard with 2-pole CBs
5. Line isolation monitors
6. Reference grounding bus
7. Equipment grounding conductors and transformer shields (if provided)
8. Enclosure

Figure 6-1. An isolated power system's equipment assembly includes transformers, line isolation monitors, overcurrent devices, a reference equipment grounding terminal bus, and an enclosure.

disconnecting pole in each isolated circuit conductor to simultaneously disconnect all power. Such isolation can be achieved by means of one or more isolation transformers, by means of generator sets or by means of electrically isolated batteries.

Motor generator sets and batteries were used in the past, but most are obsolete today since all systems are required to be listed. The most common isolated power systems used today are those supplied by equipment using isolation transformers. This equipment incorporates an isolation winding transformer, a distribution panelboard (containing circuit breakers), a line isolation monitor (LIM), and a reference grounding terminal bus, all contained within a common enclosure. A *line isolation monitor* is a test instrument designed to continually check the balanced and unbalanced impedance from each line of an isolated circuit to ground and equipped with a built-in test circuit to

exercise the alarm without adding to the leakage current hazard. The isolation windings of the transformer must have no direct electrical connection between the primary and secondary windings. **See Figure 6-1.**

The *Guide Information for Electrical Equipment* provided by Underwriters Laboratories provides some specific details about listed isolated power systems and associated equipment and accessories under the product categories KEVO, KEVX, KEWV, and KEXS. The UL Standard used to investigate and evaluate isolated power systems equipment is *UL 1047 Isolated Power Systems Equipment.*

Circuit Characteristics

If the system produces 120-volt circuits, then the phase-to-phase voltage supplying the receptacle outlets or equipment is 120 volts. There is no voltage to ground on these systems because these systems

For additional information, visit qr.njatcdb.org Item #1057

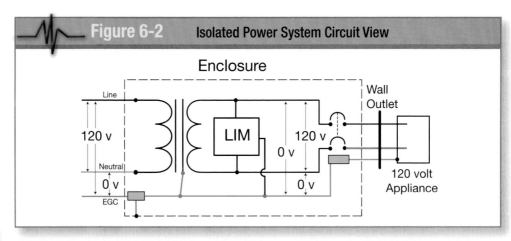

Figure 6-2. *A simplified circuit level drawing of an isolated power system from supply to outlet.*

are not grounded. Thus, the secondary circuits are considered isolated from ground. Higher voltage isolated power systems are available, but generally they are reserved for specific and dedicated equipment such as X-ray equipment or other equipment requiring greater than a 120-volt supply circuit. **See Figure 6-2.**

Where the isolated power system is a transformer-type, the primary supply circuit must not exceed 600 volts between ungrounded conductors. This circuit is required to be provided with an overcurrent protective device. This overcurrent protection device is required to protect the conductors supplying the equipment, and it must meet the size specified by the isolated power system equipment manufacturer. The secondary voltage produced by isolated power systems must not exceed 600 volts between ungrounded conductors of the system. The secondary of the isolated power systems must not be grounded, and each branch circuit supplied by such systems must be provided with an overcurrent device in series with each ungrounded conductor of the branch circuit. Therefore, for a 120-volt circuit, the overcurrent device of the breaker type or fuse type will be a 2-pole device. **See Figure 6-3.**

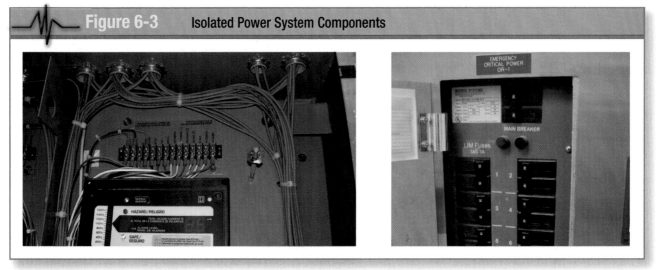

Figure 6-3. *Overcurrent protection is provided in series with each ungrounded conductor of a branch circuit supplied by an isolated power system. Notice that each circuit breaker is two-pole construction because both branch-circuit conductors are ungrounded.*

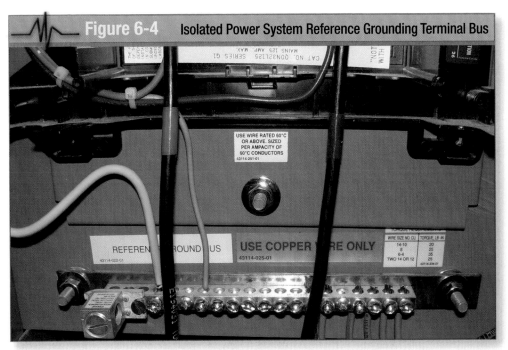

Figure 6-4. *Equipment is connected to the reference grounding terminal bus of the isolated power system.*

The circuits supplied by isolated power systems are ungrounded, but an equipment grounding conductor is required to be provided and must be connected to the reference grounding terminal bar in the listed equipment. **See Figure 6-4.**

Equipment Locations and Criteria

Most of today's isolated power system equipment is not listed for and must not be installed within a hazardous (classified) location. It is extremely important that an isolated power system be listed for the intended location and purpose. As explained in previous chapters, a hazardous location is specifically defined as one where flammable anesthetics are used, creating an explosive atmosphere. The isolated secondary circuit wiring extending into a hazardous anesthetizing location must be installed in accordance with the requirements of Article 501.10, which covers wiring methods for Class I locations. Hazardous (classified) locations are, however, rare in operating rooms (if they exist at all) these days.

Isolation Transformers

Generally, an isolated power system is limited to one operating room. However, some operating rooms include an induction room or rooms, and some operating rooms are served by higher voltage equipment such as X-ray equipment. Induction rooms are very often considered part of the operating room or rooms. The *NEC* rule about this does not focus on the entire isolated power system directly, but rather focuses on a major part of the isolated power system, the transformer. Specifically, an isolated power system transformer is only permitted to supply one operating room except under the following conditions:

* **Shared induction rooms.** An induction room is an area where anesthesia is administered prior to entering an operating room. Not all operating

Fact

UL 1047, *Isolated Power Systems Equipment*, covers isolated power systems equipment intended for installation and use in nonhazardous areas in health care facilities.

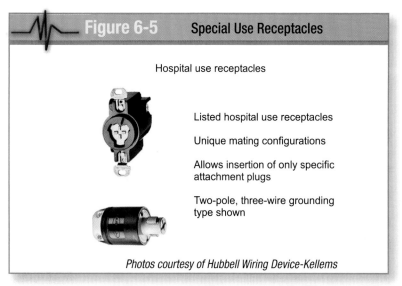

| Figure 6-5 | Special Use Receptacles |

Hospital use receptacles

Listed hospital use receptacles

Unique mating configurations

Allows insertion of only specific attachment plugs

Two-pole, three-wire grounding type shown

Photos courtesy of Hubbell Wiring Device-Kellems

Figure 6-5. Special use receptacles listed for hospital use include a special mating configuration that only allows certain attachment plugs to be inserted.

isolated power circuits feeding the induction room are permitted to be supplied from the isolated power system equipment serving any of the operating rooms served by the common induction room. [517.160(A)(4)(a)]

- *Higher voltages.* Where an isolated power system supplies equipment that operates at higher voltages such as portable X-ray equipment, the isolated power system is permitted to supply individual receptacles in more than one room if both the following are true:
  ∘ The equipment supplied requires voltages of 150 volts or higher
  ∘ The receptacles and mating plugs are not interchangeable with any receptacles supplied by the local isolated power system. [517.160(A)(4)(b) and NFPA **99**:13.4.1.2.6.6] **See Figure 6-5.**

Special purpose receptacles supplied by the IPS may be located in standalone boxes also containing monitor function lamps and equipment. **See Figure 6-6.**

rooms have separate induction areas, but when they do, the electrical code considers the induction area an extension of the operating room.

If an induction room is shared by more than one operating room, the

| Figure 6-6 | Isolated Power System Special Receptacles |

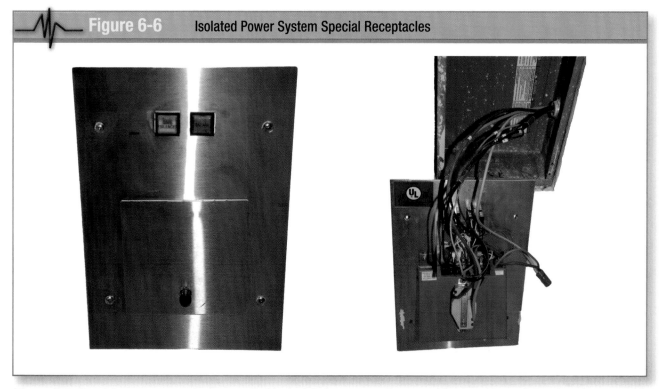

Figure 6-6. Special purpose receptacles supplied by isolated power systems provide power to equipment such as portable X-ray machines.

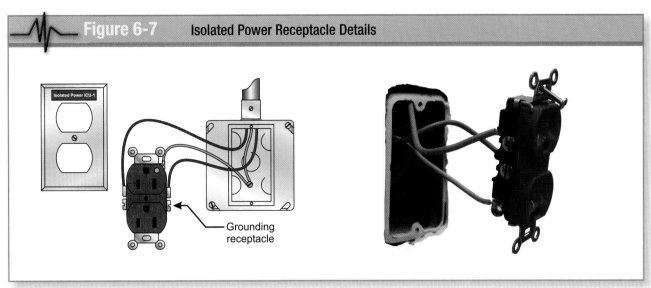

Figure 6-7 Isolated Power Receptacle Details

Grounding receptacle

Figure 6-7. *On a 125-volt, single-phase, 15- or 20-ampere receptacle, the orange conductor supplied by the isolated power system is connected to the terminal of the receptacle normally intended for the grounded system conductor.*

Conductor Identification

In 517.160(A)(5) there are specific color code identification requirements for the circuit conductors supplied by isolated power systems. Isolated power systems are available in single-phase or three-phase configurations. For single-phase units the circuit conductors shall be identified using the color orange for isolated circuit conductor No. 1, and the color brown is required for isolated circuit conductor No. 2. In addition to the specific colors (orange and brown), these conductors are also required to be identified with a distinctive colored stripe other than white, green, or gray. This required stripe must extend along the entire length of the conductor.

Where an isolated circuit conductor supplies a 125-volt, single-phase, 15- or 20-ampere receptacle, the striped orange conductor must be connected to the terminal on the receptacle that is identified for connection to the grounded circuit conductor. The grounded terminal identification means for receptacles is found in 200.10(B). **See Figure 6-7.**

Three-phase isolated power systems are not commonly used. However, where 3-phase isolated power systems are installed, they most likely are dedicated to supply specific equipment. For 3-phase systems, the third conductor is identified

by the color yellow, with a distinctive colored stripe other than white, green, or gray. **See Figure 6-8.**

The requirement for identification using orange, brown, and yellow for isolated power systems circuit conductors has been included in the *Code* for several editions. The requirement to provide orange, brown, and yellow (where used) conductors with a distinctive colored stripe was added to the 2008 edition of the *NEC* in 517.160(A)(5).

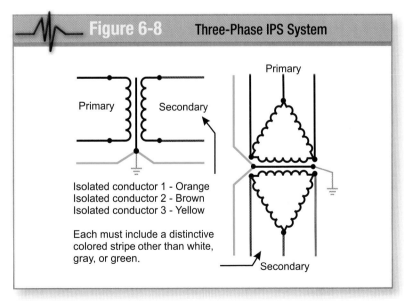

Figure 6-8 Three-Phase IPS System

Primary

Primary   Secondary

Isolated conductor 1 - Orange
Isolated conductor 2 - Brown
Isolated conductor 3 - Yellow

Each must include a distinctive colored stripe other than white, gray, or green.

Secondary

Figure 6-8. *Three-phase isolated power systems use a third, yellow conductor with a distinctive stripe.*

Fact

517.160(A)(5) indicates that where isolated circuit conductors supply 125-volt, single-phase, 15- and 20-ampere receptacles, the striped orange conductor must be connected to the white terminal on the receptacle(s).

The color code is a mandatory requirement and in the past was in conflict with the standard industry practice of using brown, orange, yellow, and gray for 480Y/277 volt systems. The *Code* has been revised in recent cycles to require that where more than one nominal voltage system exists in a building or structure, each ungrounded conductor of branch circuits or feeders must be identified by phase and system at all locations where the wiring is accessible [210.5(C) and 215.12(C)]. For the 2011 edition of the *NEC*, the required stripe along the entire length of each ungrounded conductor origination from an isolated power system does away with previous conflicts. Eventually, existing systems will transition to the required 2011 color code.

When these conditions are encountered in existing health care facilities that utilize isolated power systems equipment, care must be taken to avoid the conflict by meeting the general identification requirements in Chapter 2 with a method other than that required by 517.160(A)(5). The identification requirements in 210.5(C) and 215.12(C) recognize methods such as tagging, marking tape, color coding, or any approved means (that is, a means acceptable to the authority having jurisdiction as defined in Article 100). The means of identification chosen for the conductors must be posted at the equipment where the circuits originate. Branch-circuit identification is a requirement that relates directly to worker safety.

Limiting Leakage from Circuit Conductors

Isolated power systems provide power continuity by operating ungrounded. This means that a phase-to-ground fault condition on any of the ungrounded conductors supplied by the system will not trip an overcurrent device; it will only trigger an audible and visible signal. It is the responsibility of trained hospital staff to understand the proper response to these conditions. Isolated power systems are designed to monitor leakage current levels and provide annunciation when the circuits supplied by these system reach leakage levels that are higher than 5 milliamperes (mA). This level is very similar to the shock protection thresholds of a

Figure 6-9   **Reduced Leakage Ungrounded Conductors**

Figure 6-9. Circuit conductors supplied by an isolated power system are Type XHHW conductors. New installations require orange or brown conductors with at least one strip color other than white, green, or gray color.

ground-fault circuit interrupter. When leakage current exceeds 5 mA, the shock hazard increases. Reducing leakage current levels from line to ground reduces hazard current for patients and facility personnel. *Hazard current* is defined as the total current that would flow through a low impedance if it were connected between either isolated conductor and ground for a given set of connections in an isolated power system.

Leakage is controlled by insulation type and length of circuit. Minimizing the length of branch-circuit conductors and using conductor insulation types with a dielectric constant less than 3.5 and insulation resistance constant greater than 20,000 megohm-feet (6,100 megohm-meters) at 60°F (16°C) reduces leakage from line to ground, reducing the hazard current.

Only certain types of conductors offer an insulation that has a high enough insulation resistant constant (or low enough dielectric constant). Types THHN/THWN typically do not offer such insulation resistant values. Examples of conductor insulation types that do meet such criteria are Types XHHW and XHHW-2 insulation. **See Figure 6-9.**

The manufacturer of isolated power systems equipment usually provides specific direction for installing these systems that includes criteria such as conductor insulation types, conductor sizes, and maximum length of the circuits supplied by such systems.

Another factor that can impact the dielectric value of conductor insulation is the use of wire-pulling compounds. Wire-pulling compounds that increase the dielectric constant must not be used on the secondary conductors of the isolated power supply. All conduits and tubing used for isolated power system circuits must be cleaned and absolutely moisture free before installing wires. Generally, conduit lengths and conduit bends should also be limited; therefore, wire lubricant remains unnecessary. 110.3(B) requires that manufacturer's installation instructions be followed if circuit lengths or the number of bends are further limited. [517.160(A)(6)]

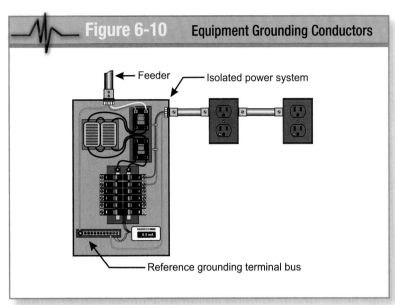

Figure 6-10. *Equipment grounding conductors are typically installed with circuits supplied by isolated power systems.*

EQUIPMENT GROUNDING REQUIREMENTS

Although the isolated power systems are ungrounded, the equipment grounding rules in Parts VI and VII of Article 250 must be satisfied. For circuits supplied by isolated power systems, an equipment grounding conductor is required on each outlet. **See Figure 6-10.**

When isolated power systems are installed, a reference to ground (the earth) is still necessary for the equipment supplied by such systems. Because the system is ungrounded, there is no solid grounding reference to the secondary side of isolated power systems. The equipment grounding requirements are satisfied by connection to an equipment grounding conductor that terminates on a reference grounding bus terminal bar located within the isolated power system equipment.

The general requirements for sizing and installation of equipment grounding conductors as presented in Article 250 are to be applied to circuits supplied by isolated power systems. For example, if the circuits supplied by an isolated power system are rated at 20 amperes, the minimum size equipment grounding conductor required by the *Code's* Table 250.122 must be no smaller than 12 AWG copper. The general requirements for equipment grounding serving patient care areas contained in Sections 517.13(A) and (B) have to be applied to these branch circuits, also. This means that redundant equipment grounding conductors are required for all branch circuits that serve patient care locations, even branch circuits supplied by isolated power systems.

The general provisions contained in Sections 300.3(B) and 250.134(B) of the *Code* require that equipment grounding conductors be installed in the same raceway, cable, or trench with the ungrounded circuit conductors. This requirement is provided for AC systems to keep the impedance levels of these systems as low as possible. By routing the equipment grounding conductors with the ungrounded circuit conductors, inductive reactance is kept low, thus reducing impedance levels in normal operation and during ground fault conditions. 517.19(F) provides a modification to this general requirement so that equipment grounding conductors associated with the ungrounded system branch-circuit conductors are permitted to be installed on the outside of the raceway containing the ungrounded circuit conductors. **See Figure 6-11.**

Although this is permitted by the minimum requirements of the *Code*, it is safer to install equipment grounding conductors in the same raceway with the power conductors to provide greater levels of protection in the event a second ground fault should develop on another ungrounded phase conductor supplied by the isolated power system. [517.19(F) and the Informational Note following] The manufacturer's installation guidelines typically provide direction for installers on the installation of equipment grounding conductors with branch circuits supplied by these systems.

LINE ISOLATION MONITORS

Because an isolated system can easily become unintentionally grounded without giving any indication to the user, a line isolation monitor must automatically and continuously check the integrity of the isolation of the system and activate an alarm should leakage current rise above the 5 milliampere threshold without interrupting the electrical service.

Function

The function of a line isolation monitor is to provide an audible and visual warning of electrical conductor insulation leakages or failures that are above acceptable levels. The normal let-go threshold for

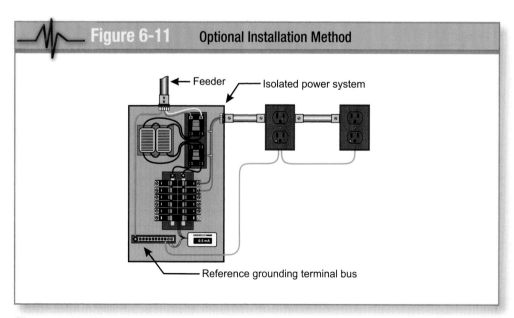

Figure 6-11   Optional Installation Method

Feeder — Isolated power system

0.5 mA

Reference grounding terminal bus

Figure 6-11. *Equipment grounding conductors installed with circuits supplied by isolated power systems may be installed on the outside of the raceway. See 517.19(F) and accompanying Informational Note.*

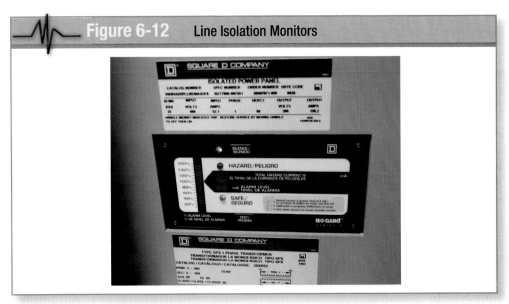

Figure 6-12. *Line isolation monitors have green and red signal lamps.*

healthy human adults (male and female) is above the 6 milliampere level. The alarm is set to function below that level; so, it does not mean there is imminent danger. It does mean, however, that the electrical system has reverted to a grounded condition (the same as in the rest of the hospital or facility), and therefore is no longer providing the same level of protection against outages that it should. The facility staff should respond to this condition by identifying the source of current leakage and removing it from the electrical system. This can mean something as simple as disconnecting equipment that has had electrical insulation failure in the past. If the insulation leakage or ground fault condition is in the electrical wiring system of the building, an electrical worker will need to correct the problem as soon as possible.

Operational Characteristics

Line isolation monitors include two signal lamps. One is green, and when lit, it indicates that the system is adequately isolated from ground. This green lamp must be conspicuously visible to persons in each area served by the isolated power system. Next to the green lamp is a red one which illuminates only when the total hazard current (consisting of possible resistive and capacitive leakage currents) from either isolated conductor to ground reaches a threshold of 5 mA under nominal line voltage conditions. At the same time the red light becomes energized, an audible warning signal (remote if desired) sounds. The monitor does not issue an alarm for a fault hazard of less than 3.7 mA or for a total hazard current of less than 5 mA unless the system is specifically designed to operate at a lower threshold value of total hazard current. *Fault hazard current* is defined as the hazard current of a given isolated system with all devices connected except the line isolation monitor. *Total hazard current* includes the line isolation monitor. **See Figure 6-12.**

For special designs using lower threshold values of hazard current, there is one exception to this requirement. Although not specifically required by the *Code*, lowering thresholds should be done only under engineering supervision. [517.160(B)(1) Exception]

Impedance Design

Line isolation monitors are required to be designed to have sufficiently low impedance so that when connected and operational, the maximum internal current levels will be detected as normal

Fact

Each isolated power system is provided with a continually operating line isolation monitor that indicates possible leakage or fault currents from either isolated conductor to ground.

conditions. The maximum internal current permitted to flow through the line isolation monitor when any point of the system is grounded is 1 mA.

An exception to this general provision recognizes line isolation monitors of the low-impedance type that allow current levels at twice the alarm threshold value for a period not exceeding 5 milliseconds. Reducing the monitor hazard current, provided this reduction results in an increased "not alarm" threshold value for the fault hazard current, will increase circuit capacity. [NEC 517.160(B)(2) Exception, and Informational note following]

Purpose and Locations of Ammeters

An ammeter calibrated in the total hazard current of the system (contribution of the fault hazard current plus monitor hazard current) must be mounted in a plainly visible place on the line isolation monitor with the "alarm on" zone at approximately the center of the scale (if the display is analog). The ammeters of older line isolation monitors were analog. Newer ones are typically digital. **See Figure 6-13.**

If the line isolation monitor is not a single unit, but a composite, it may have a sensing section cabled to a separate display panel section on which the alarm or test functions are located.

The purpose of the line isolation monitor and ammeter is to provide a visible and audible warning of current leakage levels above acceptable (safe) levels in patient care areas. Isolated power systems are not always located where visible by personnel when treating patients. For example, in an operating room, the isolated power system could be located in an adjacent location. In such a case a remote LIM is required so that staff will be able to monitor the conditions of the system while administering to the needs of the patient. **See Figure 6-14.**

Isolated power systems equipment assemblies typically include a cabinet, isolation transformer, branch-circuit panelboard, line isolation monitor, and reference grounding terminal bus within a complete assembly. Remote line isolation monitors are permitted to be connected to the assembly to satisfy the requirement that the power system be monitored from the location where the care is administered. The audible and visual indicators of the LIM are also permitted to be installed or located at a nursing station for the area served by the system. This allows facility operations staff to monitor current leakage levels.

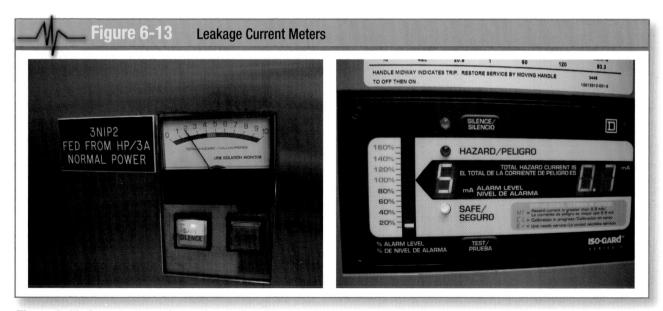

Figure 6-13 Leakage Current Meters

Figure 6-13. Ammeters may be analog or digital.

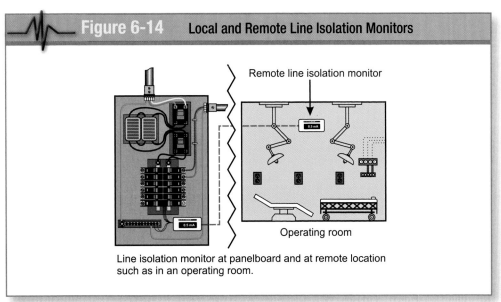

Figure 6-14   Local and Remote Line Isolation Monitors

Remote line isolation monitor

Operating room

Line isolation monitor at panelboard and at remote location such as in an operating room.

Figure 6-14. Line isolation monitors may be located at remote locations where patients are being treated, such as an operating room, to allow treatment staff to monitor the power system's status during treatment.

ACCEPTANCE TESTING (OR COMMISSIONING)

Isolated power systems are required to be performance tested when installation is completed in the facility. The following provides general guidelines and recommended procedures for acceptance testing of isolated power systems in accordance with the requirements outlined in *NFPA 99* 4.23.2.6 and 4.3.2.2.8. Manufacturers of this type of equipment provide installation, maintenance, and testing (or commissioning) criteria that meet and often exceed the minimum requirements in *NFPA 99*. Before beginning *NFPA 99* commissioning, however, each branch circuit should undergo the megohm-meter testing as required by the IPS manufacturer and outlined in Informational Note No. 2 following 517.160(A)(6).

Recommended testing procedure guidelines based on *NFPA 99* requirements are as follows:

1. Turn all branch breakers on. Unplug all portable equipment from the system under test. Turn all loads on (surgical lights, x-ray illuminators, etc.).
2. Line isolation monitors should be in the safe zone and green lights should be on. If a monitor is in alarm, turn the branch breakers off one at a time until the alarm resets. If the alarm resets, the problem is in the last branch circuit turned off. The LIM is detecting a short to ground in one of the power conductors. Do not attempt leakage readings until problem is corrected.
3. Line-to-line (X1 to X2) and line-to-ground (X1 to ground, X2 to ground) voltage readings shall be recorded (single-phase systems). If either line-to-ground reading is zero, it suggests that one isolated line conductor is actually grounded. Repeat Step 2.
4. Total Hazard Current (THC) is the total leakage current of the system with the LIM energized. Record THC from X1 to ground and X2 to ground using a calibrated meter or a high impedance AC voltmeter, using the voltage drop method prescribed by *NFPA 99*.
5. Fault Hazard Current (FHC) is the leakage current contributed by the isolated power system only. The LIM shall be removed from the circuit by unplugging the LIM. Record FHC from X1 to ground and X2 to ground.

6. Plug the LIM back in and insert a variable fault on X1 to ground until an alarm occurs. *NFPA 99* states the alarm shall not occur for a fault hazard current of less than 3.7 milliamperes, but shall alarm at 5.0 milliamperes. Repeat step for X2 ground and record as LIM alarm point.

7. Push and hold the test switch. The green light should go out and the panel meter should go up into the red zone. The red light comes on and the buzzer alarms. While holding the test switch, push the silence switch. The buzzer silences but the red light remains on. Release the test switch. See *NFPA 99* for more information regarding monthly testing by test switch.

8. The polarity of each receptacle shall be tested in compliance with *NFPA 99* and shall be corrected if demonstrated to be incorrect. [A polarity test does not apply to single-phase isolated power systems.]

9. In a given patient vicinity, the resistance between the grounding pole of one power receptacle and the grounding poles of each to the other receptacles fed from the same branch circuit shall not exceed 0.1 ohm.

Resistance measurements made between the grounding pole of a receptacle on one branch circuit and the grounding pole of a receptacle on a different branch circuit fed from the same distribution panel shall not exceed 0.5 ohms.

Resistance measurements between exposed conductive surfaces of fixed equipment and the grounding pole of a power receptacle fed from the same distribution panel shall not exceed 0.5 ohms. An exception is small wall mounted conductive surfaces not likely to become energized. [*NFPA 99*]

The resistance measured between the Patient Grounding Point and the grounding of any of the power receptacles served by the system shall not exceed 0.1 ohm.

10. Check the appearance of the conductors.

Once all of the above recommended testing and inspections are completed and the system covers are installed, the system should be re-energized and tested to verify that all equipment is operational and working properly.

Access Control Systems

An Access Control System is a system used to grant or deny access to a building, area, or room. Proper credentials are required and must be presented to grant an individual access to one of these secure areas of the building. Access control provides security for property and persons by limiting access to a specific area by use of locked doors, fences, gates, or some other form of barrier intended to block entry.

System Overview

Access Methods. Common methods of access control are human control, mechanical control, and intelligent control. One or all of these methods may be employed, depending on the level of control desired. For example, as you first enter a building you must present your ID card (credentials) to a security guard sitting at the front desk.

Once the guard approves or authorizes you to continue, you may then move further into the building. To gain access to certain floors within the building you must insert your ID card into a control slot within the elevator. Once you have reached your destination, you might have to enter a security code into a keypad to unlock a door to allow you entry into a file room.

System Selection

Human Access Control. Human Control is a form of access control that requires human intervention. The human may physically open a door or a gate or push a button to unlock a door or operate a motorized gate. A "guard" is the name given to the human providing this type of access control. Guards provide more flexibility in access control situations but also cost more to deploy than either mechanical or intelligent forms of access control.

Mechanical Access Control. Mechanical control is a form of access control that uses a mechanical device to maintain a closed area. The most common form of mechanical control is the lock and key. Another common mechanical access control device is the "mechanical keypad." A mechanical keypad operates much like a combination lock. When the correct numbers are entered in the correct sequence the lock is released.

Intelligent Access Control. Intelligent access control is an electronic system that has a microprocessor based controller or is operated from a computer system. These systems can be integrated with a building management system or a building automation system (BAS). There are several options for providing credentials to an intelligent access control system. Access cards (badges) and card readers are among the most common type of intelligent credentialing device. However, biometric devices (for example, fingerprint and retinal scanners) are becoming less expensive and more commonplace. Biometrics provides greater security than the access card, which might be "borrowed" or stolen.

System Types

Credentials. A credential is a method of authenticating an access request. A credential can be an access card, a biometric pattern, or personal identification number (PIN) that is manually entered.

A credential "reader" is a device that reads access cards or biometric patterns. Besides categorizing readers by the technology they read, they further categorize by the way that they read the credential, such as touch, insertion, and/or swipe.

There are several types of access card technologies: Barium Ferrite, bar code, magnetic stripe, proximity, Wiegand, and Smart Card.

An electronic keypad is the simplest type of electronic access control. However, when combined with other access technologies, there is a significant increase in security. These devices are known as Mixed Credential Technologies. An example would be combining a fingerprint reader or a card scanner with a keypad.

Summary

Isolated power systems are ungrounded and are required where interruption by a ground-fault circuit interrupter cannot be tolerated. Isolated power systems sense leakage current of the branch circuits supplied. When leakage current reaches predetermined levels (usually around 4 milliamperes), audible and visual warnings are required to annunciate. Isolated power systems provide critical care areas such as operating rooms and other wet procedure locations greater assurances of power continuity by operating ungrounded and by monitoring current leakage levels. A ground-fault on these systems will provide an alarm, but the circuit will continue to operate. This is important especially when patients are being operated on or are connected to life support systems that cannot afford power failure or interruption.

Review Questions

1. **Which of the following is not a characteristic of an isolated power system used in a health care facility installation?**
   a. They are required to be grounded systems.
   b. They are required to be ungrounded systems.
   c. They are required to include a line isolation monitor.
   d. The secondary windings of transformers used in isolated power systems are required to be isolated from the primary windings.

2. **The branch circuit conductors supplied by a single-phase isolated power system are required to be identified in which one of the following methods?**
   a. As approved by the authority having jurisdiction
   b. Using the colors orange and brown with a distinctive colored stripe other than white, green, or gray
   c. Using identification tags or vinyl marking tape
   d. Using number tagging means

3. **Where a branch circuit from an isolated power system panelboard supplies a 125-volt, 20-ampere, single-phase receptacle, which conductor is required to terminate on the silver terminal (terminal normally intended for the grounded conductor)?**
   a. The white conductor
   b. The brown conductor with a distinctive colored stripe other than white, green, or gray
   c. The orange conductor with a distinctive colored stripe other than white, green, or gray
   d. The green conductor

4. **Where an isolated power system is a 3-phase system supplying specific equipment, which of the following is required for branch circuit identification?**
   a. Conductor No. 1 (brown with a distinctive colored stripe other than white, green, or gray)
   b. Conductor No. 2 (orange with a distinctive colored stripe other than white, green, or gray)
   c. Conductor No. 3 (yellow with a distinctive colored stripe other than white, green, or gray)
   d. All of the above are required

Review Questions

5. **What are the principal reasons for using an isolated power system in a health care facility operating room or other critical care area?**

   a. To ensure continuity of power

   b. To monitor leakage current levels

   c. To facilitate fast overcurrent device operation in the event of a first system phase-to-ground fault condition

   d. Both (a) and (b)

6. **For isolated power systems, isolation shall be accomplished in accordance with which of the following?**

   a. By means of one or more transformers having no electrical connection between primary and secondary

   b. By means of a motor generator set

   c. By means of a suitably isolated battery

   d. Any of the above are acceptable means

7. **A(n) ? is a transformer of the multiple-winding type, with the primary and secondary windings physically separated, which inductively couples its secondary winding to the grounded feeder systems that energize its primary winding.**

   a. autotransformer

   b. separately derived system transformer

   c. isolation transformer

   d. ballast

8. **Generally, an isolated power system is permitted to serve only one operating room.**

   a. True

   b. False

9. **Wire pulling compound that will increase dielectric constants is not permitted to be used on branch circuit conductors supplied by isolated power systems.**

   a. True

   b. False

10. **The voltage supplied by a secondary of an isolated power system shall not exceed which of the following voltages?**

    a. 125 volts

    b. 240 volts

    c. 480 volts

    d. 600 volts

11. **If an electrostatic shield is provided with an isolated power system, it is required to be connected to which of the following?**

    a. A grounding electrode conductor

    b. A ground rod

    c. The reference grounding point in the isolated power system equipment

    d. The line isolation monitor

12. **The line isolation monitor functions to monitor and indicate the total hazard current in the secondary circuits supplied from the system.**

    a. True

    b. False

13. **The hazard current of a given isolated system with all devices connected except the line isolation monitor best defines which of the following?**

    a. Monitor hazard current

    b. Total hazard current

    c. Fault hazard current

    d. None of the above

14. **If an isolated power system is utilized in a critical care location or in a wet procedure location, the isolated power equipment is required to be ? .**

    a. approved

    b. identified

    c. labeled

    d. listed

15. **Where are the requirements located that may limit the number of bends in conduit or limit the maximum circuit length from the isolated power system panelboard to the outlet?**

    a. Article 517, Part II, Wiring and Protection

    b. Article 517, Part III, Essential Electrical Systems

    c. Article 517, Part IV, Isolated Power Systems

    d. Manufacturer's installation instructions

Pools and Tubs for Therapeutic Use

Health care facilities often include therapy pools and tubs for use in aquatic physical therapy, hydrotherapy, and other rehabilitation. Article 517, which covers the requirements for electrical systems in health care facilities, does not provide specific requirements for these types of equipment or electrical wiring installations. Instead, the informational note following Section 517.20 refers to the applicable requirements within Article 680, "Swimming Pools, Fountains, and Similar Installations." The applicable requirements of Article 680 are contained in Part I "General," Part II "Permanently Installed Pools," and Part VI "Pools and Tubs for Therapeutic Use."

Objectives

» Determine the shock protection requirements associated with therapeutic pools and tubs

» Describe bonding methods for therapeutic pools and tubs

» Assess grounding requirements for therapeutic pools and tubs

» Define requirements for receptacles and luminaires in or near therapeutic pools and tubs

Chapter 7

Table of Contents

GENERAL REQUIREMENTS

The wet locations and environments associated with therapeutic pools and tubs used in health care facilities must meet the general requirements of any such body of water whether within or outside of a health care facility. All such bodies of water carry specific wiring and protection requirements.

There is a major difference, however, between how recreational pools and spas are used and how they are used in the context of therapeutic or medical care. Therapeutic pools are used with people who often are incapable of saving themselves in an emergency the way people using the pools merely for recreation might be. For this reason, the electrical protections that apply to all pools, whether recreational or therapeutic, are especially important in the context of a therapeutic use.

Minimizing shock hazards and electrocution hazards in health care facilities is accomplished by establishing and maintaining little or no differences in potential within patient care locations. To accomplish this, protective techniques are used that include patient equipment grounding points and reference grounding points. A patient equipment grounding point is necessary to establish a common reference point to ground. This ensures that all electrical circuit equipment grounding conductors are connected to this point. It helps ensure that any conductive surfaces are effectively bonded together and to this point in order to minimize potential differences within the patient care vicinity.

The conductive path to the heart muscle creates increased shock hazards and concerns especially in areas where a multitude of medical equipment and appliances are routinely used for patient care. Typically, in critical care patient care locations, more medical equipment and appliances are in direct contact with patients, increasing the hazards. Where therapeutic pools and tubs are involved, the hazard is increased even more because of the presence of water and immersion of patients receiving aquatic therapeutic treatment.

The wet locations and environments associated with therapeutic pools and tubs used in health care facilities require specific electrical wiring protection methods. Electrical grounding, bonding, and installation and use of ground-fault circuit interrupters are protection methods and devices used to provide shock protection in these wet environments.

All electrical equipment installed in the water, walls, or decks of pools or similar

NEC **Definitions**

Health Care Facilities. Buildings or portions of buildings in which medical, dental, psychiatric, nursing, obstetrical, or surgical care are provided. Health care facilities include, but are not limited to, hospitals, nursing homes, limited care facilities, clinics, medical and dental offices, and ambulatory care centers, whether permanent or movable.

Packaged Therapeutic Tub or Hydrotherapeutic Tank Equipment Assembly. A factory-fabricated unit consisting of water-circulating, heating, and control equipment mounted on a common base, intended to operate a therapeutic tub or hydrotherapeutic tank. Equipment can include pumps, air blowers, heaters, lights, controls, sanitizer generators, and so forth.

Permanently Installed Swimming, Wading, Immersion, and Therapeutic Pools. Those that are constructed in the ground or partially in the ground, and all others capable of holding water in a depth greater than 1.0 m (42 in.), and all pools installed inside of a building, regardless of water depth, whether or not served by electrical circuits of any nature.

Self-Contained Therapeutic Tubs or Hydrotherapeutic Tanks. A factory-fabricated unit consisting of a therapeutic tub or hydrotherapeutic tank with all water-circulating, heating, and control equipment integral to the unit. Equipment may include pumps, air blowers, heaters, light controls, sanitizer generators, and so forth.

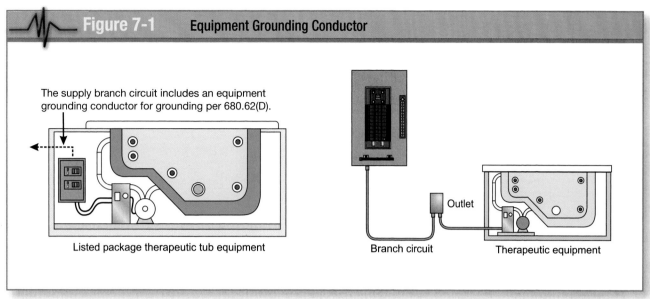

Figure 7-1 **Equipment Grounding Conductor**

The supply branch circuit includes an equipment grounding conductor for grounding per 680.62(D).

Listed package therapeutic tub equipment

Outlet

Branch circuit   Therapeutic equipment

Figure 7-1. The equipment grounding conductor must be sized according to 250.122, but no smaller than 12 AWG.

installations must be listed specifically for such use. Ground-fault circuit interrupters must be self-contained units, circuit-breaker or receptacle types, or other listed types. The following equipment must be grounded:

1. Through-wall lighting assemblies and underwater luminaires, other than those low-voltage lighting products listed for the application without a grounding conductor
2. All electrical equipment located within 5 feet (1.5 m) of the inside wall of the body of water
3. All electrical equipment associated with the recirculating system of the body of water
4. Junction boxes
5. Transformer and power supply enclosures

6. Ground-fault circuit interrupters
7. Panelboards that are not part of the service equipment and that supply any electrical equipment associated with the body of water

Cord-and-Plug Connected Equipment

Fixed or stationary equipment, other than underwater luminaires, for a permanently installed therapeutic pool is permitted to be connected with a flexible cord and plug to make it easier to remove or disconnect it for maintenance or repair. The cord-and-plug equipment must meet the following requirements:

- **Length.** The flexible cord must not be over 3 feet (900 mm) long.
- **Grounding.** The flexible cord must have a copper equipment grounding conductor sized in accordance with 250.122, but not smaller than 12 AWG. The cord must terminate in a grounding-type attachment plug.
- **Construction.** The equipment grounding conductors must be connected to a fixed metal part of the assembly. The removable part must be mounted on or bonded to the fixed metal part.

See Figure 7-1.

Fact

Electric shock can occur when the human body is in series with electric current, or in parallel with the current returning to the source. In parallel circuit contact, the victim becomes a path to ground in parallel with the equipment grounding conductor. When the human body is in series with the electrical circuit, the damage is usually most severe.

Figure 7-2   Overhead Conductor Clearances

Table 680.8 Overhead Conductor Clearances

| | | Insulated Cables, 0–750 Volts to Ground, Supported on and Cabled Together with a Solidly Grounded Bare Messenger or Solidly Grounded Neutral Conductor | | All Other Conductors Voltage to Ground | | | |
| | | | | 0 through 15 kV | | Over 15 through 50 kV | |
	Clearance Parameters	m	ft	m	ft	m	ft
A.	Clearance in any direction to the water level, edge of water surface, base of of diving platform, or permanently anchored raft	6.9	22.5	7.5	25	8.0	27
B.	Clearance in any direction to the observation stand, tower, or diving platform	4.4	14.5	5.2	17	5.5	18
C.	Horizontal limit of clearance measured from inside wall of the pool	This limit shall extend to the outer edge of the structures listed in A and B of this table but not to less than 3 m (10 ft).					

Reprinted with permission from NFPA 70-2011, *National Electrical Code®*, Copyright© 2010, National Fire Protection Association, Quincy, MA 02169. This reprinted material is not the complete and official position of the NFPA on the referenced subject, which is represented only by the standard in its entirety.

Figure 7-2. Overhead conductor clearances are given by the Code in Table 680.8.

Conductor Clearances

Indoor pools are not likely to have open spans of cables above them, but outdoor ones might. In these cases, there are rules for how far from the water these cables must be. Conductors must meet certain vertical and horizontal clearance requirements from the body of water involved. Overhead conductors must meet the following clearance requirements:

• *Power.* Service drop conductors and open overhead wiring (that is, wiring not in an enclosed raceway) above pools and similar installations must meet the minimum clearances given in Table 680.8 of the *Code.* **See Figure 7-2.**

• *Communications Systems.* Communications, radio, and television coaxial cables must be at least 10 feet (3.0 m) above pools, diving structures, and observation stands, towers, or platforms.

• *Network-Powered Broadband Communications Systems.* Network-powered broadband communications systems conductors must meet the provisions of - Table 680.8 for conductors operating at 0 to 750 volts to ground.

In all the above, where a minimum clearance from the water level is given, the measurement must be taken from the maximum water level of the body of water.

Fact

Clearances for indoor overhead luminaires, lighting outlets, and ceiling-suspended fans are covered in 680.22(B) and (C). Indoor luminaires must be totally enclosed, and fans must be identified for use beneath ceiling structures such as provided on porches or patios.

There is also a horizontal minimum that applies. The same measurements that apply overhead apply horizontally, measured from the inside wall of the pool, but at least 10 feet (3 m) in length. **See Figure 7-3.**

Electric Pool Water Heaters

All electric pool water heaters must have the heating elements subdivided into loads not exceeding 48 amperes and protected at not over 60 amperes. The ampacity of the branch-circuit conductors and the rating or setting of overcurrent protective devices must not be less than 125% of the total nameplate-rated load.

Underground Wiring Location

Pools used for therapy in a health care facility might not be installed at or below grade level. They might be on upper floors. In such cases, there is no need to worry about underground wiring locations. However, if a pool is installed at or below grade level, the following rules will apply:

- Wiring is not allowed under the pool.
- It is not allowed within the area extending 5 feet (1.5 m) horizontally from the inside wall of the pool either, unless space limitations make it necessary.
- If space limitations prevent wiring from being routed a distance of 5 feet (1.5 m) or more from the pool, such wiring is permitted if installed in complete raceway systems of rigid metal conduit, intermediate metal conduit, or a nonmetallic raceway system.
- All metal conduit must be corrosion resistant and suitable for the location.
- The raceways must be buried at minimum depths defined for each type of raceway.

Equipment Rooms and Pits

Where electrical equipment is installed in the same room as pool-water-related mechanical equipment, it must be installed in rooms or pits that have drainage that adequately prevents water accumulation during normal operation or filter maintenance.

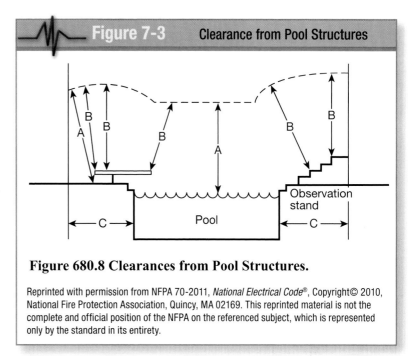

Figure 7-3   **Clearance from Pool Structures**

Figure 680.8 Clearances from Pool Structures.

Reprinted with permission from NFPA 70-2011, *National Electrical Code*®, Copyright© 2010, National Fire Protection Association, Quincy, MA 02169. This reprinted material is not the complete and official position of the NFPA on the referenced subject, which is represented only by the standard in its entirety.

Figure 7-3. Clearances from pool structures extend vertically and horizontally.

Maintenance Disconnecting Means

One or more means to simultaneously disconnect all ungrounded conductors must be provided for all utilization equipment other than lighting. Each means must be readily accessible and within sight from its equipment and located at least 5 feet (1.5 m) horizontally from the inside walls of a pool, spa, or hot tub unless separated from the open water by a permanently installed barrier that provides a 5 foot (1.5 m) reach path or greater. This horizontal distance is to be measured from the water's edge along the shortest path required to reach the disconnect.

> *NEC* **Definition**
>
> **Accessible, Readily (Readily Accessible).** Capable of being reached quickly for operation, renewal, or inspections without requiring those to whom ready access is requisite to climb over or remove obstacles or to resort to portable ladders, and so forth.

PERMANENTLY INSTALLED THERAPEUTIC POOLS

Therapeutic pools that are constructed in the ground, on the ground, or in a building in such a manner that the pool cannot be readily disassembled must comply with all relevant requirements for a permanently installed pool, regardless of location within a health care facility or not. **See Figure 7-4.**

Permanently installed swimming, wading, immersion, and therapeutic pools are pools constructed in the ground or partially in the ground, and all others capable of holding water in a depth greater than 42 inches (1.0 m), and all pools installed inside of a building, regardless of water depth, whether or not served by electrical circuits of any nature. The requirements for permanently installed pools are covered in Part II of Article 680. Not all of Part II is relevant to therapeutic pools; for instance, requirements for pools associated with single-family dwelling units [Articles 680.21(A)(4) and 680.22(A)(3)] are not relevant here and therefore are not covered.

There is one exception to the requirements of Part II when dealing with therapeutic pools, which relates to situations where totally enclosed luminaires are used. Except as noted, however, all the requirements of Part II apply to permanently installed therapeutic pools and are explained in the following pages.

Motors

For health care facilities, most pool related mechanical and electrical equipment are placed within mechanical and or electrical rooms equipped with locked doors and accessible only to maintenance staff. Wiring for branch circuits for pool-associated motors must be installed in rigid metal conduit, intermediate metal conduit, rigid polyvinyl chloride conduit, reinforced thermosetting resin conduit, or Type MC cable listed for the location. Other wiring methods and materials are permitted in specific locations or applications, but any wiring method used must contain an insulated copper equipment grounding conductor sized in accordance with 250.122, but not smaller than 12 AWG.

For motors installed on or within buildings, electrical metallic tubing is allowed. When flexible connections at or adjacent to the motor are necessary, liquidtight flexible metal or liquidtight flexible nonmetallic conduit with approved fittings is allowed. Pool-associated motors are

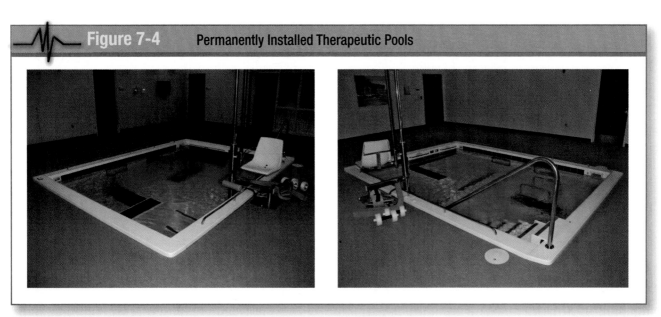

Figure 7-4 **Permanently Installed Therapeutic Pools**

Figure 7-4. *Permanently installed therapeutic pools in health care facilities follow the same rules as any permanently installed pool, in accordance with Part II of Article 680.*

Fact

Pool motors may be cord-and-plug connected but they are not required to be cord-and-plug connected. Pool motors are generally permitted to be directly connected using liquidtight flexible conduit and watertight fittings.

allowed to use cord-and-plug connections, as well. The flexible cord must not exceed 3 feet (900 mm) in length and must include a copper equipment grounding conductor sized in accordance with 250.122 but not smaller than 12 AWG and must terminate in a grounding-type attachment plug.

Double Insulated Pool Pumps

A listed cord-and-plug-connected pool pump incorporating an approved system of double insulation that provides a means for grounding only the internal and non-accessible, non-current-carrying metal parts of the pump may be connected to any wiring method suitable for the location. However, there is a requirement to extend the equipotential bonding circuit to the double insulated motor location and leave sufficient slack available should a replacement motor ever be installed that is not of the double insulated type. In addition, where there is no connection between the equipotential bonding grid and the premises equipment grounding system, this bonding conductor must also be connected to the equipment grounding conductor of the motor circuit.

GFCI Protection

Outlets supplying pool pump motors connected to single-phase, 120 volt through 240 volt branch circuits, rated 15 or 20 amperes, whether by receptacle or direct connection, must be provided with ground-fault circuit-interrupter protection for personnel.

Lighting, Receptacles, and Equipment

Receptacles in pool areas have special requirements, depending on their location and the equipment they serve. Pool equipment requires GFCI protection, whether powered from receptacles or directly wired without using receptacles. Lights, fans, and similar equipment common in pool areas require special clearances. Switching devices and outlets for such things as remote controls, fire alarms, and communication circuits all have special requirements when located in a pool area.

Receptacles

Receptacles for water-pump motors or for other loads directly related to the circulation and sanitation system must be located at least 10 feet (3.0 m) from the inside walls of the pool, or not less than 6 feet (1.83 m) from the inside walls of the pool if they meet all of the following conditions:

1. They consist of single receptacles.
2. They employ a locking configuration.
3. They are of the grounding type.
4. They have GFCI protection.

Other receptacles must be at least 6 feet (1.83 m) from the inside walls of a pool.

All 15- and 20-ampere, single-phase, 125-volt receptacles located within 20 feet (6.0 m) of the inside walls of a pool must be protected by a ground-fault circuit interrupter. Outlets supplying pool pump motors from branch circuits with short-circuit and ground-fault protection rated 15 or 20 amperes, 125 volt or 240 volt, single phase, whether by receptacle or direct connection, must be provided with ground-fault circuit-interrupter protection for personnel.

Luminaires, Lighting Outlets, and Ceiling-Suspended (Paddle) Fans

The following requirements apply to equipment commonly found in pool areas such as luminaires, lighting outlets, and ceiling fans. There is an exception for luminaires that are totally enclosed. The limitations of the first four items discussed below do not apply to situations where all luminaires are totally enclosed. [680.61]

1. *New outdoor installation clearances.* In outdoor pool areas, luminaires, lighting outlets, and ceiling-suspended (paddle) fans installed above the pool or the area

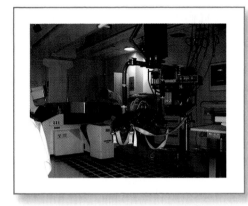

In conjunction with water immersion tubs, a Lithotripter machine is used to treat patients with kidney and bladder stones.

extending 5 feet (1.5 m) horizontally from the inside walls of the pool must be installed at a height not less than 12 feet (3.7 m) above the maximum water level of the pool.

2. ***Indoor clearances.*** For installation in indoor pool areas, the clearances are the same as for outdoor areas except as follows. If the branch circuit supplying the equipment is protected by a ground-fault circuit interrupter, the following equipment may be installed at a height not less than 7.5 feet (2.3 m) above the maximum pool water level:
   ◦ Totally enclosed luminaires
   ◦ Ceiling-suspended (paddle) fans identified for use beneath ceiling structures such as provided on porches or patios

3. ***Existing installations.*** Existing luminaires and lighting outlets located less than 5 feet (1.5 m) measured horizontally from the inside walls of a pool must not be less than 5 feet (1.5 m) above the surface of the maximum water level. They must also be rigidly attached to the existing structure, and must be protected by a ground-fault circuit interrupter.

4. ***GFCI protection in adjacent areas.*** Luminaires, lighting outlets, and ceiling-suspended (paddle) fans installed in the area extending between 5 feet (1.5 m) and 10 feet (3.0 m) horizontally from the inside walls of a pool must be protected by a ground-fault circuit interrupter unless installed at least 5 feet (1.5 m) above the maximum water level and rigidly attached to the structure adjacent to or enclosing the pool.

5. ***Cord-and-plug-connected luminaires.*** Cord-and-plug-connected luminaires installed within 16 feet (4.9 m) of any point on the water surface, measured radially, must meet all the same requirements for length, grounding, and construction discussed earlier for cord-and-plug-connected equipment:
   ◦ The flexible cord must not exceed 3 feet (900 mm) in length.
   ◦ The flexible cord must have a copper equipment grounding conductor sized in accordance with 250.122, but not smaller than 12 AWG.
   ◦ The cord must terminate in a grounding-type attachment plug.
   ◦ The equipment grounding conductors must be connected to a fixed metal part of the assembly. The removable part must be mounted on or bonded to the fixed metal part.

Switching Devices

Switching devices must be located at least 5 feet (1.5 m) horizontally from the inside walls of a pool unless separated from the pool by a solid fence, wall, or other permanent barrier; or, unless the switch is listed as being acceptable for use within 5 feet (1.5 m).

Other Outlets

Other outlets include such things as remote-control outlets, signaling outlets, fire alarm outlets, and communication circuit outlets. Such other outlets must not be less than 10 feet (3.0 m) from the inside walls of the pool. Measurements are based on the shortest path a supply cord for an appliance attached to such an outlet would follow without piercing a floor, wall, ceiling, doorway with hinged or sliding door, window opening, or other effective permanent barrier. The

10-foot distance (used as the supply cord basis) is comprised of two different practical measurements. The first distance is the distance a person located in the pool water can successfully reach outside of the pool while their body remains inside the pool (usually 4 feet) The second distance is the maximum length of a listed appliance cord for a piece of equipment plugged into an outlet, typically 6 feet. Adding the two distances gives the 10-foot distance from the pool edge to the outlet which is the *Code* minimum distance.

Underwater Luminaires

Underwater luminaires are luminaires installed below the normal water level of the pool. They may be wet-niche, dry-niche, or no-niche devices. Some requirements apply to all three, and some are unique to the particular kind of installation.

General

In general, the following requirements apply to all underwater luminaires:

- *Design and normal operation.* The design of an underwater luminaire supplied from a branch circuit either directly or by way of a transformer or power supply must be such that, where the luminaire is properly installed without a ground-fault circuit interrupter, there is no shock hazard with any likely combination of fault conditions during normal use (not relamping).
- *Transformers and Power Supplies.* Transformer and power supplies used for the supply of underwater luminaires, together with the transformer or power supply enclosure, must be listed for pool and spa use. The transformer or power supply must incorporate a transformer of the isolated winding type with an ungrounded secondary that has a grounded metal barrier between the primary and secondary windings or one that incorporates an approved system of double insulation between the primary and the secondary windings.

Fact

Added for the 2011 *NEC*, the term *Low Voltage Contact Limit* is defined as a voltage not exceeding the following values:
(1) 15 volts (RMS) for sinusoidal ac
(2) 21.2 volts peak for nonsinusoidal ac
(3) 30 volts for continuous dc
(4) 12.4 volts peak for dc that is interrupted at a rate of 10 to 200 Hz

- *Relamping.* A ground-fault circuit interrupter must be installed in the branch circuit supplying luminaires operating at more than the low voltage contact limit so that there is no shock hazard during relamping. A *low voltage contact limit* is a voltage not exceeding the following values:
(1) 15 volt (RMS) for sinusoidal AC
(2) 21.2 volts peak for nonsinusoidal AC
(3) 30 volts for continuous DC
(4) 12.4 volts peak for DC that is interrupted at the rate of 10 to 200 Hz.

  The installation of the ground-fault circuit interrupter must be such that there is no shock hazard with any likely fault-condition combination that involves a person in a conductive path from any ungrounded part of the branch circuit or the luminaire to ground.
- *Voltage limitation.* No luminaires shall be installed for operation on supply circuits over 150 volts between conductors.
- *Wall-mounted luminaires.* Luminaires mounted in walls must be installed with the top of the luminaire lens not less than 18 inches (450 mm) below the normal water level of the pool, unless the luminaire is listed and identified for use at lesser depths. No luminaire shall be installed less than 4 inches (100 mm) below the normal water level of the pool.
- *Bottom-mounted luminaires.* A luminaire facing upward must possess one of the following characteristics:
  ◦ Have the lens adequately guarded to prevent contact by any person

◦ Be listed for use without a guard
- **Dependence on submersion.** Luminaires that depend on submersion for safe operation must be inherently protected against the hazards of overheating when not submerged.
- **Compliance.** Compliance with the requirements of the *Code* is based on the use of listed underwater luminaires and by installation of a listed ground-fault circuit interrupter in the branch circuit or a listed transformer or power supply for luminaires operating at no more than the low voltage contact limit.

Wet-Niche and No-Niche Luminaires

A *wet-niche luminaire* is a luminaire intended for installation in a forming shell mounted in a pool or fountain structure where the luminaire will be completely surrounded by water. A *no-niche luminaire* is a luminaire intended for installation above or below water without a niche. Because they have closer contact with the water than a dry-niche luminaire, they have stricter requirements. Those requirements are as follows:

- **Forming shells.** Forming shells must be installed for the mounting of all wet-niche underwater luminaires and must be equipped with

provisions for conduit entries. Metal parts of the luminaire and forming shell in contact with the pool water must be of brass or other approved corrosion-resistant metal. All forming shells used with nonmetallic conduit systems, other than those that are part of a listed low-voltage lighting system not requiring grounding, shall include provisions for terminating an 8 AWG copper conductor.

- **Wiring extending directly to the forming shell.** Conduit must be installed from the forming shell to a junction box or other enclosure conforming to the requirements for junction box or other enclosure connections. These are covered below under their own topic and in the *Code* under Article 680.24. Conduit must be rigid metal, intermediate metal, liquidtight flexible nonmetallic, or rigid nonmetallic.
  1. Metal conduit must be approved and of brass or other approved corrosion-resistant metal.
  2. Where nonmetallic conduit is used, an 8 AWG insulated solid or stranded copper bonding jumper must be installed in the conduit unless a listed low-voltage lighting system not requiring grounding is used. The bonding jumper must be terminated in the forming shell, junction box or transformer enclosure, or ground-fault circuit-interrupter enclosure. The termination of the 8 AWG bonding jumper in the forming shell must be covered with, or encapsulated in, a listed potting compound to protect the connection from the possible deteriorating effect of pool water.
- **Equipment grounding provisions for cords.** For other-than-listed low-voltage lighting systems not requiring grounding, wet-niche luminaires and no-niche luminaires that are supplied by a flexible cord or cable shall have all exposed non-current-carrying metal parts grounded by an insulated copper equipment grounding conductor that is an integral part

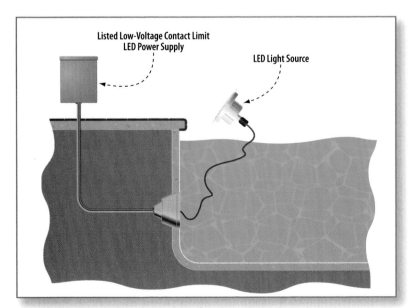

Listed Low-Voltage Contact Limit
LED Power Supply

LED Light Source

If the voltage to the lamp exceeds the low voltage contact limit, then the circuit must be protected by a GFCI.

of the cord or cable. This grounding conductor must be connected to a grounding terminal in the supply junction box, transformer enclosure, or other enclosure. The grounding conductor must not be smaller than the supply conductors and not smaller than 16 AWG.

- **Luminaire grounding terminations.** The end of the flexible-cord jacket and the flexible-cord conductor terminations within a luminaire must be covered with, or encapsulated in, a suitable potting compound to prevent the entry of water into the luminaire through the cord or its conductors. If present, the grounding connection within a luminaire must be similarly treated to protect such connection from the deteriorating effect of pool water in the event of water entry into the luminaire.

- **Luminaire bonding.** A wet-niche luminaire must be bonded to, and secured to, the forming shell by a positive locking device that ensures a low-resistance contact and requires a tool to remove the luminaire from the forming shell. Bonding is not required for luminaires that are listed for the application and have no non-current-carrying metal parts.

- **Servicing.** All wet-niche luminaires must be removable from the water for inspection, relamping, or other maintenance. The forming shell location and length of cord in the forming shell must permit personnel to place the removed luminaire on the deck or other dry location for such maintenance. The luminaire maintenance location must be accessible without entering or going into the pool water.

Dry-Niche Luminaires

A *dry-niche luminaire* is a luminaire intended for installation in the floor or wall of a pool, spa, or fountain in a niche that is sealed against the entry of water. A dry-niche luminaire must have a means for drainage and a means for accommodating one equipment-grounding conductor

Fact

An *Equipment Grounding Conductor (EGC)* is defined as the conductive path(s) installed to connect normally non–current-carrying metal parts of equipment together and to the system grounded conductor or to the grounding electrode conductor, or both.

for each conduit entry, unless it is a listed low voltage luminaire not requiring grounding.

A dry-niche luminaire does not require a junction box; and, if a junction box is used, it need not be installed according to the elevation and location requirements of a wet-niche or no-niche luminaire, which are stricter because of the closer contact these have to the water around them.

Through-Wall Lighting Assembly

A *through-wall lighting assembly* is a lighting assembly intended for installation above grade, on or through the wall of a pool, consisting of two interconnected groups of components separated by the pool wall. It must be equipped with a threaded entry or hub, or a nonmetallic hub, for the purpose of accommodating the termination of the supply conduit. A through-wall lighting assembly must meet the construction and installation requirements of wet-niche and no-niche luminaires. Where connection to a forming shell is specified, the connection must be to the conduit termination point.

Branch-Circuit Wiring

Branch-circuit wiring on the supply side of enclosures and junction boxes connected to conduits run to wet-niche and no-niche luminaires, and the field wiring compartments of dry-niche luminaires shall be installed using rigid metal conduit, intermediate metal conduit, liquidtight flexible nonmetallic conduit, rigid polyvinyl chloride conduit, or reinforced thermosetting resin conduit. Where installed on buildings, electrical metallic tubing is permitted, and where installed within buildings, electrical nonmetallic tubing, Type MC cable, electrical metallic tubing, or Type AC cable is permitted. In all cases, an insulated equipment grounding conductor sized in accordance with Table 250.122 of the *Code*, but not less than 12 AWG, is required.

There is one exception to this. Where connecting to transformers for pool lights, liquidtight flexible metal conduit or liquidtight flexible nonmetallic conduit is permitted. The length must not exceed 6 feet (1.8 m) for any one length or 10 feet (3.0 m) in total length used.

Equipment grounding. Other than listed low voltage luminaires not requiring grounding, all through-wall lighting assemblies, wet-niche, dry-niche, or no-niche luminaires must be connected to an insulated copper equipment grounding conductor installed with the circuit conductors. The equipment grounding conductor must be sized in accordance with Table 250.122 of the *Code*, but not smaller than 12 AWG, with the following exception. An equipment grounding conductor between the wiring chamber of the secondary winding of a transformer and a junction box must be sized in accordance with the overcurrent device in this circuit.

The equipment grounding conductor must be installed without joint or splice, except as follows:

- If more than one underwater luminaire is supplied by the same branch circuit, the equipment grounding conductor installed between the junction boxes, transformer enclosures, or other enclosure in the supply circuit to wet-niche luminaires, or between the field-wiring compartments of dry-niche luminaires, is permitted to be terminated on grounding terminals.

> ⚕ **Fact**
>
> Revised for the 2011 *NEC*, a *Bonding Conductor or Jumper* is defined as a reliable conductor to ensure the required electrical conductivity between metal parts required to be electrically connected.

- If the underwater luminaire is supplied from a transformer, ground-fault circuit interrupter, clock-operated switch, or a manual snap switch that is located between the panelboard and a junction box connected to the conduit that extends directly to the underwater luminaire, the equipment grounding conductor may terminate on grounding terminals on the transformer, ground-fault circuit interrupter, clock-operated switch enclosure, or an outlet box used to enclose a snap switch.

Conductors. Conductors on the load side of a ground-fault circuit interrupter or a transformer must not occupy raceways, boxes, or enclosures containing other conductors unless one of the following conditions applies:

1. The other conductors are protected by ground-fault circuit interrupters.
2. The other conductors are grounding conductors.
3. The other conductors are supply conductors to a feed-through-type ground-fault circuit interrupter.
4. Ground-fault circuit interrupters are permitted in a panelboard that contains circuits protected by other than ground-fault circuit interrupters.

Junction Boxes and Other Electrical Enclosures

Junction boxes and other electrical enclosures connected to conduits that extend directly to a forming shell or mounting bracket of a no-niche luminaire have special requirements.

Construction

A junction box, as well as other electrical enclosures, connected to a conduit extending directly to a forming shell or mounting bracket of a no-niche luminaire must be listed for pool use and must possess the following characteristics:

1. They must be equipped with threaded entries or hubs or a nonmetallic hub.
2. They must be comprised of copper, brass, suitable plastic, or other approved corrosion-resistant material.

3. They must be provided with electrical continuity between every connected metal conduit and the grounding terminals by means of copper, brass, or other approved corrosion-resistant metal that is integral with the box.

Other electrical enclosures include enclosures for transformers, ground-fault circuit interrupters, and similar devices. These have the same requirements for their construction, with one addition. In addition to the requirements they share with junction boxes, these other enclosures require an approved seal, such as duct seal at the conduit connection, that prevents circulation of air between the conduit and the enclosure.

Installation

Where the luminaire operates over the low voltage contact limit, a junction box (and any other electrical enclosure) must be installed in a location that meets the following specifications for vertical and horizontal spacing:

- **Vertical spacing.** The junction box must be located not less than 4 inches (100 mm), measured from the inside of the bottom of the box, above the ground level, or pool deck, or not less than 8 inches (200 mm) above the maximum pool water level, whichever provides the greater elevation.
- **Horizontal spacing.** The junction box must be located not less than 4 feet (1.2 m) from the inside wall of the pool, unless separated from the pool by a solid fence, wall, or other permanent barrier.

Where the luminaire operates at the low voltage contact limit, the junction box location may be flush with the deck if both of the following conditions are met:

1. An approved potting compound is used to fill the box to prevent the entrance of moisture.
2. The flush deck box is located not less than 4 feet (1.2 m) from the inside wall of the pool.

No such horizontal installation is allowed for enclosures other than junction boxes. Other enclosure installations must meet the requirements for vertical and horizontal spacing.

Protection

Junction boxes and enclosures mounted above the grade of the finished walkway around the pool must not be located in the walkway unless afforded additional protection, such as by location under diving boards, adjacent to fixed structures, and the like.

Grounding Terminals

Junction boxes, transformer and power supply enclosures, and ground-fault circuit-interrupter enclosures connected to a conduit that extends directly to a forming shell or mounting bracket of a no-niche luminaire must be provided with a number of grounding terminals that shall be no fewer than one more than the number of conduit entries.

Strain Relief

The termination of a flexible cord of an underwater luminaire within a junction box, transformer or power supply enclosure, ground-fault circuit interrupter, or other enclosure must be provided with a strain relief.

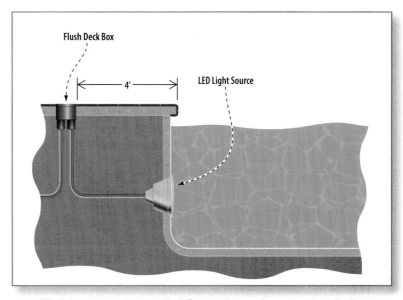

An LED light source is supplied from a low-voltage contact limit power supply.

Grounding

The equipment grounding conductor terminals of a junction box, transformer enclosure, or other enclosure in the supply circuit to a wet-niche or no-niche luminaire and the field-wiring chamber of a dry-niche luminaire must be connected to the equipment grounding terminal of the panelboard. This terminal shall be directly connected to the panelboard enclosure.

Feeders

The following requirements apply to any feeder on the supply side of panelboards supplying branch circuits for pool equipment and on the load side of the service equipment or the source of a separately derived system.

Wiring Methods

Feeders must be installed in rigid metal conduit, or intermediate metal conduit. The following wiring methods shall be permitted if not subject to physical damage:

1. Liquidtight flexible nonmetallic conduit
2. Rigid polyvinyl chloride conduit
3. Reinforced thermosetting resin conduit
4. Electrical metallic tubing where installed on or within a building

For additional information, visit qr.njatcdb.org Item #1058

5. Electrical nonmetallic tubing where installed within a building
6. Type MC cable where installed within a building and not subject to a corrosive environment

Aluminum conduit shall not be permitted in the pool area where subject to corrosion.

There is an exception to the above. See Article 680.25(A)(1) Exception. An existing feeder between an existing remote panelboard and service equipment may run in flexible metal conduit or an approved cable assembly that includes an equipment grounding conductor within its outer sheath. The equipment grounding conductor must comply with 250.24(A)(5).

Grounding

An equipment grounding conductor must be installed with the feeder conductors between the grounding terminal of the pool equipment panelboard and the grounding terminal of the applicable service equipment or source of a separately derived system. This equipment grounding conductor must be insulated, except under two conditions:

1. If it is the equipment grounding conductor associated with an existing feeder, as described above

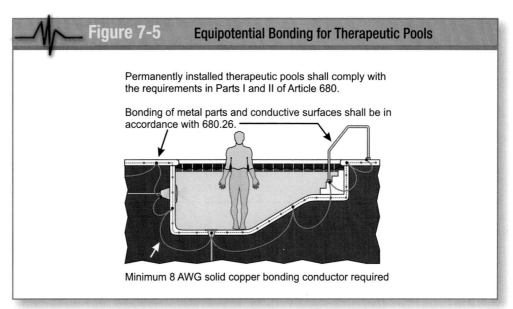

Figure 7-5. Equipotential bonding at a permanently installed therapeutic pool reduces voltage gradients throughout the pool area.

[680.25(A) Exception], it need not be insulated.

2. If the equipment grounding conductor is associated with feeders to separate buildings that do not utilize an insulated equipment grounding conductor, it need not be insulated. A feeder to a separate building or structure is allowed to supply pool equipment branch circuits, or feeders supplying pool equipment branch circuits, if the grounding arrangements in the separate building meet the requirements in Article 250.32(B)(1). Where installed in other-than-existing feeders covered under Article 680.25(A), Exception, a separate equipment grounding conductor must be an insulated conductor.

The conductor must be sized in accordance with Section 250.122, but must not be smaller than 12 AWG. On separately derived systems, this conductor must be sized in accordance with 250.30(A)(8), but must not be smaller than 8 AWG.

Equipotential Bonding

Equipotential bonding is required to reduce voltage gradients in the pool area. It applies to the several bonded parts of the pool structure as well as to the pool water itself. Equipotential bonding is not related to overcurrent device operation, which is the primary performance function of an equipment grounding conductor. Rather, the purpose of equipotential bonding is to minimize voltage gradients in the therapeutic pool area. The *Code* requires an 8 AWG solid copper bonding conductor for this purpose. **See Figure 7-5.**

The 8 AWG or larger solid copper bonding conductor is not required to be extended or attached to any remote panelboard, service equipment, or electrode because it is not installed for purposes served by an equipment grounding conductor.

Bonded Parts
Parts to be bonded include:
1. Conductive pool shells
2. Perimeter surfaces

Fact

Equipotential bonding is just one of the many shock protection techniques used for wiring in and around pool areas.

3. Metallic components
4. Underwater lighting
5. Metal fittings
6. Electrical equipment
7. Metal wiring and equipment

Each of these is to be bonded using solid copper conductors (insulated, covered, or bare) not smaller than 8 AWG or with rigid metal conduit of brass or other identified corrosion-resistant metal. Connections to the bonded parts must be made in accordance with Article 250.8. An 8 AWG or larger solid copper bonding conductor provided to reduce voltage gradients in the pool area need not be extended or attached to remote panelboards, service equipment, or electrodes.

Conductive pool shells. Poured concrete, pneumatically applied or sprayed concrete, and concrete block with painted or plastered coatings are all considered conductive materials due to water permeability and porosity. Vinyl liners and fiberglass composite shells are considered to be non-conductive materials.

Bonding to conductive pool shells must be done as follows:
1. Unencapsulated structural reinforcing steel must be bonded together by steel tie wires or the equivalent. Where structural reinforcing steel is encapsulated in a nonconductive compound, a copper conductor grid must be installed.
2. If a copper conductor grid is used, it must:
   a. Be constructed of minimum 8 AWG bare solid copper conductors bonded to each other at all points of crossing. The bonding must be in accordance with 250.8 or other approved means.
   b. Conform to the contour of the pool

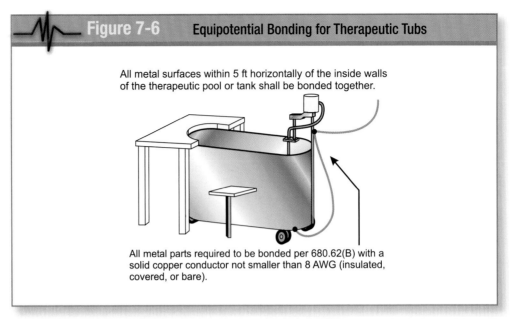

Figure 7-6 Equipotential Bonding for Therapeutic Tubs

All metal surfaces within 5 ft horizontally of the inside walls of the therapeutic pool or tank shall be bonded together.

All metal parts required to be bonded per 680.62(B) with a solid copper conductor not smaller than 8 AWG (insulated, covered, or bare).

Figure 7-6. *Bonding metal parts of the therapeutic tank or tub is required to accomplish equipotential bonding.*

c. Be arranged in a 12-inch by 12-inch (300-mm by 300-mm) network of conductors in a uniformly spaced perpendicular grid pattern with a tolerance of 4 inches (100 mm)

d. Be secured within or under the pool no more than 6 inches (150 mm) from the outer contour of the pool shell

Perimeter surfaces. The perimeter surface extends for 3 feet (1 m) horizontally beyond the inside walls of the pool and includes unpaved surfaces as well as poured concrete surfaces and other types of paving. Perimeter surfaces less than 3 feet separated by a permanent wall or building 5 feet in height or more shall require equipotential bonding on the side of the permanent wall or building. Bonding to perimeter surfaces must be provided as specified above using either structural reinforcing steel or a copper grid. There must be a minimum of four attachment points to the pool's reinforcing steel or copper conductor grid from a minimum of four points uniformly spaced around the perimeter of the pool. This does not apply to nonconductive pool shells.

If unencapsulated structural reinforcing steel is used, it must be bonded as

described above, using steel tie wires or their equivalent.

If structural reinforcing steel is not available or is encapsulated in a nonconductive compound, a copper conductor must be used and the following requirements must be met:

1. At least one minimum 8 AWG bare solid copper conductor must be provided.
2. The conductors must follow the contour of the perimeter surface.
3. Only listed splices are allowed.
4. The required conductor must be 18 to 24 inches (450 to 600 mm) from the inside walls of the pool.
5. The required conductor must be secured within or under the perimeter surface 4 to 6 inches (100 to 150 mm) below the subgrade.

Metallic components. All metallic parts of the pool structure, including reinforcing metal not addressed already, must be bonded, except for reinforcing steel that is encapsulated with a nonconductive compound. **See Figure 7-6.**

Underwater lighting. All metal forming shells and mounting brackets of no-niche luminaires must be bonded. Listed low-voltage lighting systems with nonmetallic forming shells do not require bonding; only the ones with metal forming shells do.

Metal fittings. All metal fittings within or attached to the pool structure must be bonded with the following exception. Isolated parts that are not over 4 inches (100 mm) in any dimension and do not penetrate into the pool structure more than 1 inch (25 mm) are not required to be bonded.

Electrical equipment. Metal parts of electrical equipment associated with the pool water circulating system, including pump motors and metal parts of equipment associated with pool covers, including electric motors, must be bonded. There is an exception to this for metal parts of listed equipment that incorporates an approved system of double insulation.

Metal parts of listed equipment incorporating an approved system of double insulation shall not be bonded. This exception affects the following two situations:

1. **Double-insulated water pump motors.** Where a double-insulated water pump motor is installed under the provisions of this rule, a solid 8 AWG copper conductor of sufficient length to make a bonding connection to a replacement motor must be extended from the bonding grid to an accessible point in the vicinity of the pool pump motor. Where there is no connection between the pool bonding grid and the equipment grounding system for the premises, this bonding conductor must be connected to the equipment grounding conductor of the motor circuit.
2. **Pool water heaters.** For pool water heaters rated at more than 50 amperes and having specific instructions regarding bonding and

grounding, only those parts designated to be bonded must be bonded and only those parts designated to be grounded must be grounded.

Fixed metal parts. All fixed metal parts must be bonded including, but not limited to, metal-sheathed cables and raceways, metal piping, metal awnings, metal fences, and metal door and window frames, with the following exceptions:

1. Those separated from the pool by a permanent barrier that prevents contact by a person need not be bonded.
2. Those greater than 5 feet (1.5 m) horizontally from the inside walls of the pool need not be bonded.
3. Those greater than 12 feet (3.7 m) measured vertically above the maximum water level of the pool, or as measured vertically above any observation stands, towers, or platforms, or any diving structures, need not be bonded.

Pool Water

An intentional bond of a minimum conductive surface area of 9 square inches (5,800 mm^2) must be installed in contact with the pool water. This bond may consist of the parts required to be bonded in 680.26(B), "Bonded Parts," as described in the previous section. The handrail of a metal ladder or the ring of a pool light is one example of compliance with this requirement.

Specialized Pool Equipment

Specialized equipment includes such things as underwater audio equipment, electric pool covers, and deck heating. Though more common in recreational facilities, such amenities are not uncommon in health care uses.

Underwater Audio Equipment

All underwater audio equipment must be listed for such use. It includes:

- **Speakers.** Each speaker must be mounted in an approved metal forming shell, the front of which is enclosed by a captive metal screen, or equivalent, that is bonded to, and

Fact

Double insulated is a shock protection technique applied to electrical products. It is comprised of two insulation systems (basic and supplementary) that are physically separated.

Figure 7-7	Various Therapeutic Tubs

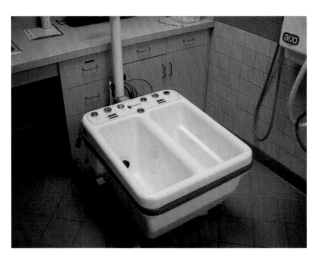

Figure 7-7. Therapeutic tubs may be stationary or movable, and can come in a variety of sizes.

secured to, the forming shell by a positive locking device. The locking device must ensure a low-resistance contact and require a tool to open it for installation or servicing of the speaker. The forming shell must be installed in a recess in the wall or floor of the pool.

- **Wiring.** Rigid metal conduit of brass or other identified corrosion-resistant metal, liquidtight flexible nonmetallic conduit (LFNC-B), rigid polyvinyl chloride conduit, or reinforced thermosetting resin conduit must extend from the forming shell to a listed junction box or other enclosure as described in Article 680.24, "Junction Boxes and Electrical Enclosures."

  Where rigid polyvinyl chloride conduit, reinforced thermosetting resin conduit, or liquidtight flexible nonmetallic conduit is used, an 8 AWG insulated solid or stranded copper bonding jumper must be installed in the conduit. The bonding jumper must be terminated in the forming shell and the junction box. The termination of the 8 AWG bonding jumper in the forming shell must be covered with, or encapsulated in, a listed potting compound to protect the connection from the possible deteriorating effect of pool water.

- **Forming shell and metal screen.** The forming shell and metal screen must be of brass or other approved corrosion-resistant metal. All forming shells must include provisions for terminating an 8 AWG copper conductor.

Electrically Operated Pool Covers

The electric motors, controllers, and wiring for an electrically operated pool cover must be located at least 5 feet (1.5 m) from the inside wall of the pool unless separated from the pool by a wall, cover, or other permanent barrier. Electric motors installed below grade level must be of the totally enclosed type. The device that controls the operation of the motor for an electrically operated pool cover must be located in a place that gives the operator full view of the pool.

The motor and controller must be connected to a circuit protected by a ground-fault circuit interrupter.

Deck Area Heating

The following requirements apply all pool deck areas, including a covered pool, where electrically operated comfort heating units are installed within 20 feet (6.0 m) of the inside wall of the pool.

1. Unit heaters must be rigidly mounted to the structure and be of the totally enclosed or guarded type. They must not be mounted over the pool or within the area extending 5 feet (1.5 m) horizontally from the inside walls of a pool.

2. Permanently wired radiant heaters must be suitably guarded and securely fastened to their mounting devices. Heaters must not be installed over a pool or within the area extending 5 feet (1.5 m) horizontally from the inside walls of the pool and must be mounted at least 12 feet (3.7 m) vertically above the pool deck unless otherwise approved.

3. Radiant heating cables may not be embedded in or below the deck.

THERAPEUTIC TUBS (HYDROTHERAPEUTIC TANKS)

Therapeutic tubs, used for the submersion and treatment of patients, that are not easily moved from one place to another in normal use or that are fastened or otherwise secured at a specific location, including associate piping systems, must meet requirements in the following areas:

1. Protection from shock
2. Bonding
3. Bonding methods
4. Grounding
5. Receptacles
6. Luminaires

See Figure 7-7.

Protection from Shock

Except as otherwise provided below, the outlet that supplies a self-contained therapeutic tub or hydrotherapeutic tank, a packaged therapeutic tub or hydrotherapeutic tank, or a field-assembled therapeutic tub or hydrotherapeutic tank must be protected by a ground-fault circuit interrupter. GFCI protection can be provided by a receptacle or outlet device or by a circuit breaker that incorporates ground-fault circuit-interrupter protection. **See Figure 7-8.**

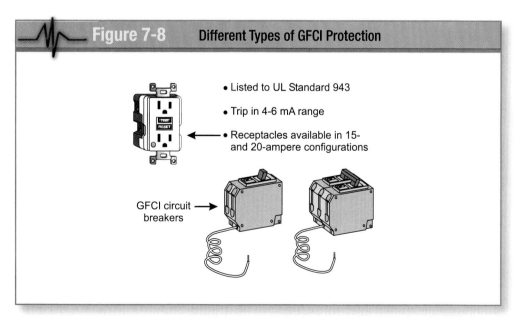

Figure 7-8   Different Types of GFCI Protection

- Listed to UL Standard 943
- Trip in 4-6 mA range
- Receptacles available in 15- and 20-ampere configurations

GFCI circuit breakers

Figure 7-8. GFCI protection may come from receptacles or circuit breakers.

Figure 7-9   Integral GFCI Protection

GFCI protection is required to be provided unless it is included with a listed packaged therapeutic tub or hydrotherapeutic tank assemby.

Integral ground-fault circuit-interrupter protection

Listed package therapeutic tub unit assembly

Figure 7-9. GFCI protection may be an integral part of the hydrotherapeutic equipment.

Exceptions are as follows:
1. If so marked, a listed self-contained unit or listed packaged equipment assembly that includes integral ground-fault circuit-interrupter protection for all electrical parts within the unit or assembly (pumps, air blowers, heaters, lights, controls, sanitizer generators, wiring, and so

forth) may omit the additional GFCI protection. **See Figure 7-9.**
2. A therapeutic tub or hydrotherapeutic tank rated 3 phase or rated over 250 volts or with a heater load of more than 50 amperes does not require its supply to be protected by a ground-fault circuit interrupter.

Bonding

The following parts must be bonded together:
1. All metal fittings within or attached to the tub structure
2. Metal parts of electrical equipment associated with the tub water circulating system, including pump motors
3. Metal-sheathed cables and raceways and metal piping that are within 5 feet (1.5 m) of the inside walls of the tub and not separated from the tub by a permanent barrier
4. All metal surfaces that are within 5 feet (1.5 m) of the inside walls of the tub and not separated from the tub area by a permanent barrier
5. Electrical devices and controls that are not associated with the therapeutic tubs and located within 5 feet (1.5 m) from such units

Small conductive surfaces not likely to become energized are exempt from these bonding requirements. Examples of such small conductive surfaces include air and water jets, drain fittings not connected to metallic piping, towel bars, mirror frames, and similar nonelectrical equipment not connected to metal framing.

Bonding Methods

All metal parts required to be bonded must be bonded in one of the following ways:
• The interconnection of threaded metal piping and fittings
• Metal-to-metal mounting on a common frame or base
• Connections by suitable metal clamps
• By the provisions of a solid copper bonding jumper, insulated, covered, or bare, not smaller than 8 AWG
See Figure 7-10.

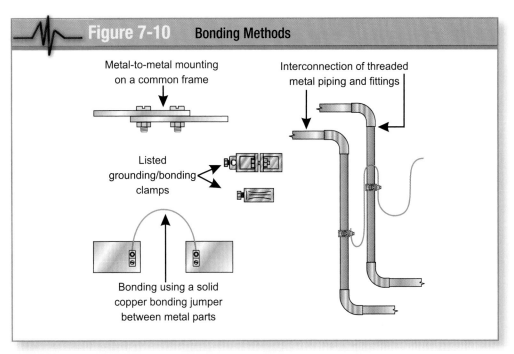

Figure 7-10   Bonding Methods

Metal-to-metal mounting on a common frame

Interconnection of threaded metal piping and fittings

Listed grounding/bonding clamps

Bonding using a solid copper bonding jumper between metal parts

Figure 7-10. There are several acceptable methods for bonding metal parts of a therapeutic tub.

Grounding

The following equipment must be connected to the equipment grounding conductor:

- All fixed or stationary electrical equipment located within 5 feet (1.5 m) of the inside wall of the tub
- All fixed or stationary electrical equipment associated with the circulating system of the tub
- Portable therapeutic appliances that fall under the rules of Article 250.114

Receptacles

All receptacles within 6 feet (1.83 m) of a therapeutic tub must be protected by a ground-fault circuit interrupter. **See Figure 7-11.**

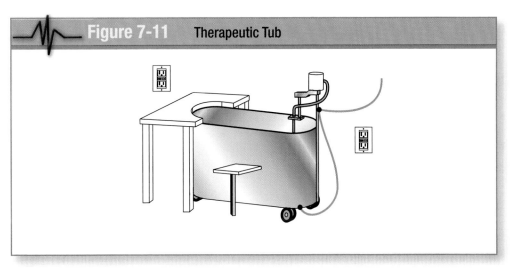

Figure 7-11   Therapeutic Tub

Figure 7-11. Receptacles within 6 feet (1.83 m) of a therapeutic tub or hydrotherapeutic tank must be protectd by a GFCI.

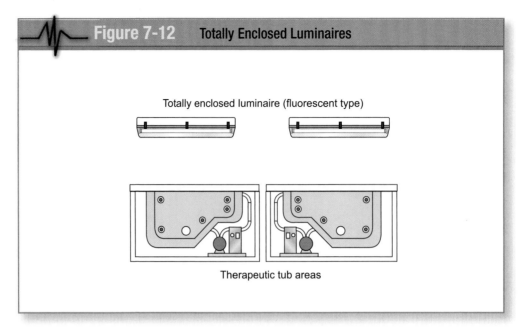

Figure 7-12   Totally Enclosed Luminaires

Totally enclosed luminaire (fluorescent type)

Therapeutic tub areas

Figure 7-12. Luminaires in therapeutic tub areas must be totally enclosed.

Luminaires

All luminaires used in therapeutic tub areas must be of the totally enclosed type. The *Code* does not place specific vertical height rules or horizontal distances for the luminaires used in these areas. Therefore, they must be totally enclosed. **See Figure 7-12.**

Fact

See 250.118 for a list of acceptable equipment grounding conductors.

Radio Frequency Identification

Radio Frequency Identification (RFID) is a type of automatic identification system. The purpose of an RFID system is to enable data to be transmitted by a portable device, called a "tag," which is read by an RFID reader and processed according to the needs of a particular application. The data transmitted by the tag may provide identification, location information, or specifics about the product tagged, such as price, color, date of purchase, etc.

System Overview

In a typical RFID system, individual objects are equipped with a small, inexpensive tag which contains a transponder with a digital memory chip that is given a unique electronic product code or personal ID. The interrogator, an antenna packaged with a transceiver and decoder, emits a signal activating the RFID tag so it can read and write data to it. When an RFID tag passes through the electromagnetic zone, it detects the reader's activation signal. The reader decodes the data encoded in the tag's integrated circuit (silicon chip) and the data is passed to the host computer for processing.

RFID tags come in a wide variety of shapes and sizes. Some tags are easy to spot, such as the hard plastic anti-theft tags attached to merchandise in stores. Animal tracking tags which are implanted beneath the skin of family pets or endangered species are no bigger than a small section of pencil lead. Even smaller tags have been developed to be embedded within the fibers of a national currency.

System Selection

RFID technology in health care facilities is being put to work in different systems, such as asset management, patient and baby tracking, patient-bedside care, and improving Operating Room (OR) procedure safety.

System Types

For managing health care facility assets, an RFID tag is attached to each piece of facility equipment. Strategically locating readers throughout the area of operations where the equipment is used, an Asset Management Software Program then keeps track of where the equipment is located and whether or not it is in use. Consumable goods such as syringes, gauze, sponges, and towels may be tracked using RFID as well as tracking the location and amounts of medications used by the health care facility (Medication Verification System).

Hospitals' operating rooms use RFID to keep track of all of the items that enter and leave the OR that are used for surgery. The purpose is to reduce or eliminate unintentionally leaving foreign objects in a patient after surgery. The RFID system automatically provides an item count by directly matching the unique identifier on each tagged item both entering into and then out of the OR prior to and after the surgery. An RFID detecting wand may be used to scan the patient as an additional safety measure.

RFID systems provide a less expensive solution for active RF transmitter based Patient Wandering and Infant Tracking systems. RF transmitters systems require a wrist bracelet or ankle bracelet with an active (battery powered) transmitter that is sending a signal 24 hours a day. RFID tags do not require any batteries and can be encoded with patient data, such as required medication or other records.

Summary

Water and wet locations present additional and increased hazards and concerns for patient and staff safety in health care facilities. Therapeutic tubs, tanks, and pools requiring electrical power must be installed according to specific rules in Chapter 6 of the *NEC*. The requirements for health care therapeutic tubs and pools are provided in Part VI of Article 680. The Informational Note following Section 517.20 provides the correlation to the rules in Article 680 that pertain to this type of equipment. Equipment grounding is required for these therapeutic tubs and pools whether they are permanently installed or movable. Specific bonding requirements apply to this type of equipment to assure equipotential between metal parts and other conductive surfaces and the branch circuits supplying therapeutic pools and tubs are required to be protected by Class A ground-fault circuit-interrupter protection.

Review Questions

1. **Which part(s) of Article 680 apply to pools and tubs or tanks for therapeutic use in health care facilities?**

   a. Part I only

   b. Part I and VI

   c. Part VI only

   d. All parts apply to these types of installations

2. **Which part(s) of Article 680 apply to therapeutic pools or tubs that are constructed in the ground in a building in such a manner that they cannot be readily disassembled?**

   a. Parts I and II

   b. Part VI only

   c. Part I only

   d. None of the above

3. **Therapeutic tubs or hydrotherapeutic tanks that are supplied with a 480 volt, 3-phase circuit are not required to be protected with which of the following?**

   a. An overcurrent protective device

   b. An equipment grounding conductor

   c. Ground-fault circuit interrupter protection

   d. All of the above are required

4. **Where a bonding jumper of the wire type is used for bonding metal parts of therapeutic tubs (hydrotherapeutic tanks), what are the required characteristic(s) for this bonding jumper?**

   a. It must not be smaller than 8 AWG

   b. It is required to be copper

   c. It is required to be solid

   d. All of the above

Review Questions

5. **Receptacles installed within 6 feet (1.83 m) of a therapeutic tub are required to meet which of the following requirements?**

   a. They must be provided with weatherproof enclosures.

   b. They are required to be in nonmetallic enclosures.

   c. They are required to be installed using not larger than 12 AWG insulated copper conductors.

   d. They are required to be protected by a ground-fault circuit interrupter.

6. **Luminaires used in therapeutic tub areas are required to be of the __?__ type.**

   a. incandescent

   b. fluorescent

   c. metal halide

   d. totally enclosed

7. **Where a listed self-contained therapeutic tub assembly is marked to indicate that it includes integral ground-fault circuit-interrupter protection for all electrical parts in the assembly, additional ground-fault circuit-interrupter protection is not required.**

   a. True

   b. False

8. **What is the primary purpose of bonding metal parts that are associated with therapeutic pools or tubs in health care facilities?**

   a. Minimize differences of potential between metal parts

   b. Facilitate overcurrent device operation

   c. Establish an effective connection to the earth

   d. All of the above

9. **Portable therapeutic appliances are covered by the requirements contained in which of the following?**

   a. Parts II and III of Article 422

   b. Parts VI and I of Article 680

   c. Neither (a) or (b)

   d. Both (a) and (b)

10. **A packaged therapeutic tub or hydrotherapeutic tank equipment assembly consists of which of the following?**

    a. A factory-fabricated unit consisting of water-circulating, heating, and control equipment mounted on a common base

    b. Pumps, air blowers, heaters, and lights

    c. Controls, sanitizer generators, and so forth

    d. All of the above

11. **A __?__ is a factory-fabricated unit consisting of a therapeutic tub or hydrotherapeutic tank with all water-circulating, heating, and control equipment integral to the unit. Its equipment may include pumps, air blowers, heaters, light controls, sanitizer generators, and so forth.**

    a. A hot tub or spa

    b. A self-contained therapeutic tub or hydrotherapeutic tank

    c. A packaged therapeutic tub or hydrotherapeutic tank equipment assembly

    d. A hydromassage bathtub

12. **Which of the following *NEC* articles provides the requirements for therapeutic pools and hydrotherapeutic tanks used in health care facilities?**

    a. Article 517

    b. Article 422

    c. Article 682

    d. Article 680

13. **Which of the following outlets supplying a permanently installed therapeutic pool pump motor requires ground-fault circuit-interrupter protection for personnel?**

    a. 120 volt, 30 amps, single phase

    b. 208 volt, 20 amps, single phase

    c. 208 volt, 20 amps, three phase

    d. 480 volt, 15 amps, single phase

Working in Operational Facilities

Working in an operational health care facility requires special procedures. The electrical worker must schedule work so as not to disturb the facility's normal functions. Work must be done in a way that does not interfere with staff doing their jobs. There will be areas of the facility that require special access permission because of the records, supplies, or special equipment kept in them. One must learn how to work around patients who require privacy and who may have sensitivities to debris, noise, or other work-related by-products. Finally, the electrical worker needs to provide for his or her own personal safety in an environment with special hazards.

Objectives

» Summarize the special procedures in scheduling and performing work in a functioning health care facility

» Explain issues surrounding patient confidentiality

» Identify and explain the various types of special personal protective equipment commonly provided when working in an operational health care facility

Chapter 8

Table of Contents

WORKING PROCEDURES

Working in an operational health care facility carries unique risks to both residents of the facility and the electrical worker. The contractor will perform a risk assessment prior to beginning work and will plan the work to minimize risks. First, the type of construction will be defined to determine if it consists of any of the following:

- Inspection and non-invasive activities, such as removal of ceiling tiles for visual inspection
- Small scale, short duration activities that generate little dust, such as installing telephone cables
- Work that generates moderate to high levels of dust, such as sanding or new wall construction; any work that cannot be completed within a single work shift
- Major demolition, such as replacing an entire cabling system or adding a new addition

See Figure 8-1.

After identifying the type of construction involved, the areas affected will be classified according to risk group. Areas

For additional information, visit qr.njatcdb.org Item #1059

⁄\/\— **Figure 8-1**	**Classifying Types of Construction**
Construction Classifications	
Category	**Activities**
TYPE A	**Inspection and non-invasive activities** Activities include, but are not limited to: • removal of ceiling tiles for visual inspection limited to 1 tile per 50 square feet • painting (but not sanding) • wallcovering, electrical trim work, minor plumbing, and activities which do not generate dust or require cutting of walls or access to ceilings other than for visual inspection
TYPE B	**Small scale, short duration activities which create minimal dust** Activities include, but are not limited to: • installation of telephone and computer cabling • access to chase spaces • cutting of walls or ceiling where dust migration can be controlled
TYPE C	**Work that generates a moderate to high level of dust or requires demolition or removal of any fixed building components or assemblies** Activities include, but are not limited to: • sanding of walls for painting or wall covering • removal of floorcoverings, ceiling tiles and casework • new wall construction • minor duct work or electrical work above ceilings • major cabling activities • any activity which cannot be completed within a single workshift
TYPE D	**Major demolition and construction projects** Activities include, but are not limited to: • activities which require consecutive work shifts • requires heavy demolition or removal of a complete cabling system • new construction

Figure 8-1. *Classifying the type of construction will help determine the safe working procedures needed.*

Figure 8-2 — Areas of Work and Their Associated Levels of Risk

Low Risk	Medium Risk	High Risk	Highest Risk
• Office Areas	• Cardiology • Echocardiography • Endoscopy • Nuclear Medicine • Physical Therapy • Radiology/MRI • Respiratory Therapy	• CCU • Emergency Room • Labor & Delivery • Laboratories (Specimen) • Newborn Nursery • Outpatient Surgery • Pediatrics • Pharmacy • Post Anesthesia Care Unit • Surgical Units	• Any Area Caring for Immuno-compromised Patients • Burn Unit • Cardiac Cath Lab • Central Sterile Supply • Intensive Care Units • Medical Unit • Negative Pressure Isolation Rooms • Oncology • Operating Rooms, Including C-Section Rooms

Figure 8-2. Classifying the affected resident population according to risk will help determine the necessary safe working practices.

such as offices and reception rooms are likely to be considered areas where the people in them are at low risk for infection or irritation by construction debris or dust. Other areas will have occupants with higher sensitivities. Medium risk areas include physical therapy rooms. Higher risk areas include outpatient surgery, and even higher risk areas would include operating rooms. **See Figure 8-2.**

Depending on the type of construction, and the level of risk group involved, different precautions will be needed. Correlating the risk group with the type of construction defines a classification system for determining appropriate work procedures. **See Figure 8-3.**

Figure 8-3 — Associating Level of Risk with Type of Construction

Category/Risk Correlation				
	TYPE A Construction	**TYPE B Construction**	**TYPE C Construction**	**TYPE D Construction**
LOW Risk Group	I	II	II	III/IV
MEDIUM Risk Group	I	II	III	IV
HIGH Risk Group	I	II	III/IV	IV
HIGHEST Risk Group	II	III/IV	III/IV	IV

Figure 8-3. Correlating the risk group with the construction type establishes a precaution class which will then determine the appropriate working procedures to manage risks.

Based on the risk/construction type classification, precautions are defined for work performed during construction and for post-construction cleanup. **See Figure 8-4.**

Example 1: Reception Area Water Leak
The reception area of the health care facility is normally considered a low risk area. If a leak were to cause water to flow into the ceiling spaces of a reception area, damaging tile and perhaps compromising computer cabling in the space, the repairs might be minor or they might be extensive depending on the extent of the damage. Simply checking the ceiling space would be a Type A construction activity. Removing tiles and replacing cables might be a Type B activity. If the water damage is severe, replacing damaged drywall, rugs, and flooring and working on repairs for more than a day or two would turn the activity into a Type C or even Type D activity.

Combining the low risk area with a construction activity type (A, B, C, or D) determines the construction classification. A combination of low risk area with Activity A results in a Class I project. Combining the low risk area with Activity B or C results in a Class II project. Finally, combining the low risk area with Activity D results in a Class III project, bordering on Class IV.

Based on the project classification (I, II, III, or IV), certain precautions are required.

Example 2: Minor ICU Ceiling Damage
If a water leak spread to the ceiling of an intensive care unit (ICU), but did not appear serious, one might just lift a couple of tiles to look into the ceiling space to evaluate the situation. This is a case where the area is automatically a very high risk area. Even though the activity is very minor, the area is so high risk that the project is considered a Class II project. Active measures need to be taken to prevent dust from moving from the ceiling space to the ICU atmosphere.

After completing inspection, surfaces might need to be wiped with disinfectant. The area might need to be wet mopped or vacuumed with a HEPA filtered vacuum. Other measures might be needed to prevent infection or contamination from entering or exiting the area.

Example 3: Cables Behind Nursery Wall
A newborn nursery in a health care facility is considered a high risk area (not as high a risk as the ICU, but significantly higher than a reception area). Cutting through a wall in such an area to replace cabling would certainly result in dust and debris, representing a Type B construction activity and therefore a Class II project. If the cabling became a major activity requiring more than a day to complete, the construction activity would be considered a Type C activity and the project a Class III project.

The precautions required for the project class would include isolating the HVAC system during construction and vacuuming with HEPA filtered vacuums after construction, as well as the many other precautions to be taken for Class III projects.

Other Considerations

Each project and task is unique and must be evaluated individually. The above guidelines and examples are only that – guidelines and examples of the kinds of things to consider. Other considerations that bear on determining proper work procedures include, but are by no means limited to, the following:

1. *Areas surrounding the project area.* Rooms above or below or next to the work area must be assessed to determine the potential impact of the project on those areas.
2. *The specific site of activity.* For example, patient rooms, medication room, record keeping areas, imaging diagnostic rooms where magnetic forces or radiation exposure may pose a hazard.
3. *Issues related to ventilation, plumbing, electrical outages.*
4. *Containment measures.* For instance, solid wall barriers or HEPA filtration

Figure 8-4 Work Procedures Related to Class of Construction

Precautions During and After Construction

	During Construction	After Construction
CLASS I	1. Execute work by methods to minimize raising dust from construction operations. 2. Immediately replace a ceiling tile displaced for visual inspection.	
CLASS II	1. Provide active means to prevent airborne dust from dispersing into atmosphere. 2. Water mist work surfaces to control dust while cutting. 3. Seal unused doors with duct tape. 4. Block off and seal air vents. 5. Place dust mat at entrance and exit of work area. 6. Remove or isolate HVAC system in areas where work is being performed.	1. Wipe work surfaces with disinfectant. 2. Contain construction waste before transport in tightly covered containers. 3. Wet mop and/or vacuum with HEPA filtered vacuum before leaving work area. 4. Remove isolation of HVAC system in areas where work is being performed.
CLASS III	1. Remove or isolate HVAC system in area where work is being done to prevent contamination of duct system. 2. Complete all critical barriers, i.e., sheetrock, plywood, plastic, to seal area from non-work area or implement control cube method (cart with plastic covering and sealed connection to work site with HEPA vacuum for vacuuming prior to exit) before construction begins. 3. Maintain negative air pressure within work site utilizing HEPA equipped air filtration units. 4. Contain construction waste before transport in tightly covered containers. 5. Cover transport receptacles or carts. Tape covering unless solid lid.	1. Do not remove barriers from work area until completed project is inspected by the owner's Safety Department and Infection Control Department and thoroughly cleaned by the owner's Environmental Services Department. 2. Remove barrier materials carefully to minimize spreading of dirt and debris associated with construction. 3. Vacuum work area with HEPA filtered vacuums. 4. Wet mop area with disinfectant. 5. Remove isolation of HVAC system in areas where work is being performed.
CLASS IV	1. Isolate HVAC system in area where work is being done to prevent contamination of duct system. 2. Complete all critical barriers, i.e., sheetrock, plywood, plastic, to seal area from non-work area or implement control cube method (cart with plastic covering and sealed connection to work site with HEPA vacuum for vacuuming prior to exit) before construction begins. 3. Maintain negative air pressure within work site utilizing HEPA equipped air filtration units. 4. Seal holes, pipes, conduits, and punctures appropriately. 5. Construct anteroom and require all personnel to pass through this room so they can be vacuumed using a HEPA vacuum cleaner before leaving work site, or they can wear cloth or paper coveralls that are removed each time they leave the work site. 6. All personnel entering work site are required to wear shoe covers. Shoe covers must be changed each time the worker exits the work area. 7. Do not remove barriers from work area until completed project is inspected by the owner's Safety Department and Infection Control Department and thoroughly cleaned by the owner's Environmental Services Department.	1. Remove barrier material carefully to minimize spreading of dirt and debris associated with construction. 2. Contain construction waste before transport in tightly covered containers. 3. Cover transport receptacles or carts. Tape covering unless solid lid. 4. Vacuum work area with HEPA filtered vacuums. 5. Wet mop area with disinfectant. 6. Remove isolation of HVAC system in areas where work is being performed.

Figure 8-4. Working procedures are determined by precaution class, which is in turn determined by correlating the construction type and risk group.

5. *Potential water damage risk.* Is there a risk due to compromising structural integrity of wall, ceiling, or roof?

6. *Restraints on normal work hours.* Can or will the work be done during non-patient care hours?

7. *The need for isolation/negative airflow rooms.*

8. *The required number and type of hand-washing sinks.* Does the infection control staff agree with the minimum number of sinks for the project? Verify against AIA Guidelines for types and area.

9. *Utility rooms.* Does the infection control staff agree with the plans relative to clean and soiled utility rooms?

Before working on a project in a health care facility, the electrical worker will be briefed on the special procedures that apply to that project.

UNIQUE CONSTRUCTION ISSUES

Patients in hospitals and other health care facilities may be more sensitive to things that do not affect healthy people. For instance, spores of environmental fungi are everywhere in the environment and have little effect on healthy people. However, they may cause disease in patients with compromised immune systems. Since the spores are everywhere, they might be carried in construction dust. Containment procedures will be very important in a health care setting. [JD. Siegel, E. Rhinehart, M. Jackson, L Chiarello, and the healthcare Infection Control Practices Advisory committee, *2007 Guideline for Isolation Precautions: Preventing Transmission of Infectious Agents in Healthcare Settings.*

Fact

Many workers are unaware of chemicals that create potential hazards in their work environment, making them more vulnerable to exposure and injury.

Knowing Hospital Communication Codes

One of the unique things about working in a hospital is the need to know some of the codes used by the hospital to summon emergency teams. If a team is about to come running down the corridor a person is working in, they need to be able to pass. The worker needs to know to get out of the way ahead of time. If the code is specific to an area of the hospital, and the worker understands it, he or she can decide if it applies to the area where they are working. For instance, in some hospitals, a Code 99 is also a Code Blue, meaning a patient is in some form of arrest. Such a code may be floor-specific since patients will be assigned to rooms in the hospital.

An M.E.T. (medical emergency team) call, on the other hand, means someone is in arrest somewhere in the hospital. It could be a patient or a visitor, and the emergency could be located in any part of the hospital. The emergency team will be coming from all different directions. So, a worker in any part of the hospital will need to clear the work area quickly.

Working in Small Increments

The need to clear a work area quickly means that work must sometimes be performed in smaller increments than usual. For instance, instead of removing all the tiles in a ceiling to replace or install conduit, it may be necessary to remove only one tile at a time. This will mean working with shorter pieces of conduit, using more couplings, etc., in order to close off work quickly.

Working in small increments also means cleaning up frequently. If there is debris that has accumulated during the work, it cannot be left lying around while emergency teams race through it.

Keeping a materials cart in a nearby storage room, instead of in the actual workspace, will help with rapid cleanup. Having only one ladder in the work area to move instead of two, for instance, will help the process.

If it is necessary to leave the area for a short time (a break or lunch, for instance), the area needs to be cleaned or contained.

Infection Control Construction Permit					
				Permit No:	
Location of Construction:			Project Start Date:		
Project Coordinator:			Estimated Duration:		
Contractor Performing Work			Permit Expiration Date:		
Supervisor:			Telephone:		
YES	NO	CONSTRUCTION ACTIVITY	YES	NO	INFECTION CONTROL RISK GROUP
		TYPE A: Inspection, non-invasive activity			GROUP 1: Low Risk
		TYPE B: Small scale, short duration, moderate to high levels			GROUP 2: Medium Risk
		TYPE C: Activity generates moderate to high levels of dust, requires greater than 1 work shift for completion			GROUP 3: Medium/High Risk
		TYPE D: Major duration and construction activities requiring consecutive work shifts			GROUP 4: Highest Risk

CLASS I
1. Execute work by methods to minimize raising dust from construction operations.
2. Immediately replace any ceiling tile displaced for visual inspection.

3. Minor Demolition for Remodeling

CLASS II
1. Provides active means to prevent air-borne dust from dispersing into atmosphere
2. Water mist work surfaces to control dust while cutting.
3. Seal unused doors with duct tape.
4. Block off and seal air vents.
5. Wipe surfaces with disinfectant.

6. Contain construction waste before transport in tightly covered containers.
7. Wet mop and/or vacuum with HEPA filtered vacuum before leaving work area.
8. Place dust mat at entrance and exit of work area.
9. Remove or isolate HVAC system in areas where work is being performed.

CLASS III

Date

Initial

1. Obtain infection control permit before construction begins.
2. Isolate HVAC system in area where work is being done to prevent contamination of the duct system.
3. Complete all critical barriers or implement control cube method before construction begins.
4. Maintain negative air pressure within work site utilizing HEPA equipped air filtration units.
5. Do not remove barriers from work area until complete project is thoroughly cleaned by Env. Services Dept.

6. Vacuum work with HEPA filtered vacuums.
7. Wet mop with disinfectant
8. Remove barrier materials carefully to minimize spreading of dirt and debris associated with construction.
9. Contain construction waste before transport in tightly covered containers.
10. Cover transport receptacles or carts. Tape covering.
11. Remove or isolate HVAC system in areas where work is being performed.

CLASS IV

Date

Initial

1. Obtain infection control permit before construction begins.
2. Isolate HVAC system in area where work is being done to prevent contamination of duct system.
3. Complete all critical barriers or implement control cube method before construction begins.
4. Maintain negative air pressure within work site utilizing HEPA equipped air filtration units.
5. Seal holes, pipes, conduits, and punctures appropriately.
6. Construct anteroom and require all personnel to pass through this room so they can be vacuumed using a HEPA vacuum cleaner before leaving work site or they can wear cloth or paper coveralls that are removed each time they leave the work site.

7. All personnel entering work site are required to wear shoe covers
8. Do not remove barriers from work area until completed project is thoroughly cleaned by the Environmental Service Dept.
9. Vacuum work area with HEPA filtered vacuums.
10. Wet mop with disinfectant.
11. Remove barrier materials carefully to minimize spreading of dirt and debris associated with construction.
12. Contain construction waste before transport in tightly covered containers.
13. Cover transport receptacles or carts. Tape covering.
14. Remove or isolate HVAC system in areas where is being done.

Additional Requirements:

	Exceptions/Additions to this permit
Date Initials	Date Initials
	are noted by attached memoranda
Permit Request By:	Permit Authorized By:
Date:	Date:

Health care facilities frequently require special building permits, based on the nature of the work to be performed. One such permit is the Infection Control Construction Permit shown here.

Fact

Hazardous and toxic substances are defined as those chemicals present in the workplace which are capable of causing harm. In this definition, the term *chemicals* includes dusts, mixtures, and common materials such as paints, fuels, and solvents.

If work will continue beyond a single shift or day, the area needs to be cleaned and cleared of ladders, extension cords, and any other construction materials that might interfere with health care operations.

Containment

As already mentioned, it is important to contain any dust or other construction debris within the construction area so it does not carry disease to other areas of the hospital or simply create an irritant for patients with decreased immunities and more sensitive allergy systems.

If the work has caused more mess than can easily be swept up, the facility's housekeeping department must be notified and the worker will have to wait until they respond before leaving.

This is a sensitive issue with regulations changing every day. It may eventually be mandatory that the worker be confined to a rolling tent, zippered shut to contain any type of dust from becoming airborne.

MRI Room Safety

The magnet in an MRI is always on even when not doing a procedure; therefore, any time work is being done in these rooms special precautions need to be taken. The danger is that the electrical worker may get too close to the actual magnet. The strength depends on the amount of metal and its proximity to the magnet. When the MRI magnet grabs any kind of metal, it will pull it directly to it. Tools can fly out of a worker's hands, damaging the equipment. Worse, if someone is between the magnet and the flying tool, he or she can be impaled. If a worker has any kind of metal in his or her body (for example, screws, pins, or even a small piece of metal in the skin) the magnet is powerful enough to pull this metal out of the body if he or she comes too close to the magnet.

The MRI room can be a very dangerous place if strict precautions are not observed. Metal objects can become dangerous projectiles. Paper clips, pens, keys, scissors, and any other small objects can be pulled out of pockets and off the body without warning, at which point they fly toward the opening of the magnet at very high speeds, posing a threat to everyone in the room. Credit cards, bank cards, and anything else with magnetic encoding will be erased by most MRI systems.

The employer shall establish an emergency evacuation plan. Any emergency action plan shall be in writing and shall cover those designated actions that employers and employees must take to ensure employee safety from fire and other emergencies. This includes designating a rally point on the job site and accounting for all employees. The employers emergency evacuation plan must be in concert with the plan of heath care facility and approved by health care administration.

A contractor's emergency evacuation plan must be in concert with the plan of the health care facility.

Lead Walls

All X-ray walls include some degree of shielding, the thickness and composition of which will vary depending on the amount of radiation contained in the room and the uses of adjacent rooms.

Newer equipment emits less radiation; therefore, newer construction may not contain the same level of shielding as older. Radiation tends to dissipate over space, so the nearness of an X-ray machine to a wall will affect how much shielding is needed. The direction of the beam will affect which areas of a room require most shielding.

Radiation exposure risks are cumulative. Therefore, repeated exposure at low doses may still be dangerous. If a room next to an X-ray room is only used for storage and does not often have people in it, it may not require as much shielding as a room that is generally occupied.

All materials provide some degree of shielding. Concrete, concrete block, and even multiple layers of drywall may be sufficient in some cases. However, in many cases lead is used to shield an X-ray room. The determination of how much shielding is needed and what materials to use will be made by a qualified professional. The electrical worker needs to know how to deal with whatever the resulting construction contains.

For instance, a normal X-ray room will often have lead lined drywall (standard 5/8" drywall with 1/8" lead glued to the backing). A room used to administer radiation therapy may be constructed of solid lead blocks furred out and covered with drywall. Any electrical device boxes in such walls must be covered with 1/8" sheet lead glued to the boxes. This is true for any type of box installed in a wall in such a radiation area.

PATIENT PRIVACY

Patients in a health care facility have a right to privacy. This right is protected by law under the Health Insurance Portability and Accountability Act (HIPAA) of 1996. This law applies to the health care facility, which is required to

> **Fact**
>
> Both nuclear radiation and X-ray medical shielding installed during hospital construction and remodel include lead products such as lead brick, lead-lined plywood and gypsum board, and lead sheets. Other medical shielding may include shielded wood and metal doors, shielded glass, and view ports.

establish rules with its contractors and other entities known as "business associates" that will protect the privacy of the facility's patients.

The law does not necessarily impose penalties on people who only accidentally learn something. For instance, a doctor talking about a patient in the elevator is putting him or herself at risk, but the person who overhears the conversation may not be subject to the law. However, if that person represents a contractor to the facility, at the very least, that person puts his employer's relationship with the facility at risk if he or she discloses anything overheard. Even though the electrical worker may not be legally liable for violating someone's privacy (assuming the information was only accidentally learned and not deliberately stolen), it is still a good work practice to carefully honor the privacy requirements of the health care facility for which one works.

If the worker is exposed to an infectious disease, a patient's right to privacy may become an issue in determining how the exposure occurred. Under the HIPAA rules, health care organizations may disclose protected health information that they believe is necessary to prevent or lessen a serious and imminent threat to a person or the public, when such disclosure is made to someone they believe can prevent or lessen the threat (including the target of the threat). [*OCR Privacy Brief: Summary of the HIPAA Privacy Rule*, U.S. Department of Health and Human Services, 2003, p. 6.] One needs to be sensitive to the patient's right to privacy, but workers also have a right to protect themselves.

 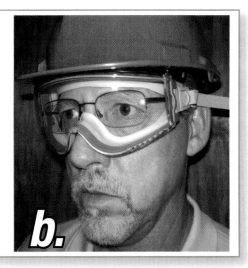

(a) Face and eye protection must be kept clean, in good repair, and free of defects. (b) When eye protection is required to be worn by employees whose vision requires the use of corrective lenses, the lenses must be protected by goggles.

HEALTH CARE FACILITY PPE

In general, safety is maintained through a hierarchy of training and administrative controls, engineering controls, work practice controls, and personal protective equipment.

The general requirements of OSHA laws regarding PPE are covered in OSHA 1910.132. The health care administration, together with the contractor(s), assesses the hazards involved and decides if PPE is necessary. If the hazards are present or likely to be present, the administration, together with the contractor(s), selects the types of PPE that will protect the affected employees from those hazards. Before any exposure occurs, each exposed worker must be fit tested for the assigned PPE. In addition, each exposed worker must be trained in when and how to use it as well as how to care for the assigned PPE. Guidelines for hazard assessment and PPE selection are found in Part 1910, Subpart I, Appendix B of the OSHA regulations.

In a health care facility, the electrical worker is always at risk of coming into contact with anything a sick patient has brought with them. If working in a patient room, even if the patient is not in the room, one may still be exposed to infection. If working in a patient room that housekeeping has decided not to clean until after the work is done, a worker may even be exposed to bodily fluids.

The principal PPE an electrical worker is likely to use will be latex gloves. The principal work practice one will follow is hand washing, with anti-bacterial soap, followed by hospital grade hand sanitizer. Another major caution is that workers avoid touching their faces while working until after sanitizing their hands.

It may be worthwhile to know some of the other procedures and PPE that health care facilities recommend for their own people.

Within the health care environment, personal protective equipment is defined by OSHA as "specialized clothing or equipment worn by an employee for protection against infectious materials." This is in addition to any normal PPE the electrical worker might already use. OSHA specifies the circumstances under which PPE is required. From a medical point of view, the actual PPE, how to use it, and when to use it are covered by the Centers for Disease Control (CDC). [*Guidance for the Selection and Use of Personal Protective Equipment (PPE) in Healthcare Settings*, CDC, June 29, 2004.] Depending on the situation, an electrical worker might be required to wear some of the same personal protective equipment that a health care worker wears. Knowing something about these items, the proper way to put them on and take them off, and the proper way to dispose of them will be useful.

Such items include:

- Gloves – protect hands
- Gowns/aprons – protect skin and/or clothing
- Masks – protect mouth/nose
- Respirators – protect respiratory tract from airborne infectious agents
- Goggles – protect eyes
- Face shields – protect face, mouth, nose, and eyes

Decisions regarding when and which type of PPE should be worn are determined by CDC recommendations for Standard Precautions and Expanded Isolation Precautions. Standard Precautions is an outgrowth of Universal Precautions. Universal Precautions was first recommended in 1987 to prevent the transmission of blood-borne pathogens to health care personnel. In 1996, the application of the concept was expanded and renamed "Standard Precautions." "Standard Precautions" is intended to prevent the transmission of common infectious agents to health care personnel, patients, and visitors in any health care setting.

Whether PPE is needed, and if so, which type, is determined by the type of interaction with the patient and the degree of contact that can be reasonably anticipated and by whether the patient has been placed on isolation precautions, such as Contact or Droplet Precautions or Airborne Infection Isolation.

The decision to require any of these is largely the employer's decision. The choice will depend not only on the best way to protect the worker, but also the best way to protect any patients who might be affected by the worker's activities. Each item of PPE has its variations and needs to be selected based on the activity involved.

Gloves

Gloves, for instance, carry several considerations. They are chosen based on the following:

- Purpose – patient care, environmental services, other
- Glove material – vinyl, latex, nitrile, other
- Sterile or nonsterile
- One or two pair
- Single use or reusable

Environmental services personnel often wear reusable heavy duty gloves made of latex or nitrile to work with caustic disinfectants when cleaning environmental surfaces. However, they sometimes use patient care gloves, too. The electrical worker might be expected to work with these, also. These kinds of gloves are not dielectric, however. They protect against infection, not electrical shock, and are therefore not a substitute for normal electrical protective gloves.

Gloves protect against contact with infectious materials. However, once contaminated, gloves can become a means for spreading infectious materials to oneself, patients, or environmental surfaces. Therefore, the way gloves are used can influence the risk of disease transmission in a health care setting. The following are the most important procedures of glove use:

- ***Work from clean to dirty.*** This is a basic principle of infection control. In this instance it refers to touching clean surfaces before touching dirty or heavily contaminated areas.
- ***Limit opportunities for "touch contamination."*** After touching a potentially contaminated surface, do not touch yourself, for example, scratch your nose, adjust your glasses, or touch your face. This is one example of "touch contamination" that can potentially expose oneself to infectious agents. Think

Fact

29 CFR Parts 1910 and 1926 provide both general requirements for things such as *Criteria for Personal Protective Equipment,* as well as specific requirements for subjects such as *Bloodborne Pathogens.*

For additional information, visit qr.njatcdb.org Item #1087

about environmental surfaces too and avoid unnecessarily touching them with contaminated gloves. Surfaces such as light switches and door and cabinet knobs can become contaminated if touched by soiled gloves.

- *Change gloves as needed.* If gloves become torn or heavily soiled and additional tasks must be performed, then change the gloves before starting the next task. Always change gloves after use, and discard them in the nearest appropriate receptacle. Patient care gloves should never be washed and used again. Washing gloves does not necessarily make them safe for reuse; it may not be possible to eliminate all microorganisms and washing can make the gloves more prone to tearing or leaking.

Gowns/Coveralls

There are three factors that influence the selection of a gown or coverall as PPE.

1. First is the purpose of use. Isolation gowns are generally the preferred PPE for clothing but aprons occasionally are used where limited contamination is anticipated. If contamination of the arms can be anticipated, a gown should be selected. Gowns should fully cover the torso, fit comfortably over the body, and have long sleeves that fit snugly at the wrist. Among electrical workers, a common high-density polyethelene fiber coverall is frequently used. These fabrics are breathable, but prevent water from passing through. These can come with hoods for full body coverage. They protect against particles as small as 0.5 microns, even when the fabric has been abraded.

2. Second are the material properties of the gown. Isolation gowns are made either of cotton or a spun synthetic material that dictate whether they can be laundered and reused or must be disposed. This coverall can be a reusable or a disposable item, depending on the style chosen. The material is a single layer, rather than a laminated material. This means that scratches do not destroy the protective barrier of the coverall.

3. The third factor concerns patient risks and whether a clean, rather than sterile, gown can be used. Clean gowns are generally used for isolation. Only a health care worker, such as a surgeon, would be expected to need a sterile gown.

Masks

A combination of PPE types is available to protect all or parts of the face from contact with potentially infectious material. The selection of facial PPE is determined by the isolation precautions required for the patient and/or the nature of the patient contact.

Masks should fully cover the nose and mouth. Masks should fit snugly over the nose and mouth. For this reason, masks that have a flexible nose piece and can be secured to the head with string ties or elastic are preferable.

Goggles

Goggles provide barrier protection for the eyes; personal prescription lenses do not provide optimal eye protection and should not be used as a substitute for goggles. Goggles should fit snugly over and around the eyes or personal

Hazardous communication training includes how to use personal protective equipment (PPE) and material safety data sheets (MSDS) information.

prescription lenses. Goggles with antifog features will help maintain clarity of vision and are usually considered as a "required" optional feature.

Face Shields

When skin protection, in addition to mouth, nose, and eye protection, is needed or desired, a face shield can be used as a substitute to wearing a mask or goggles. The face shield should cover the forehead, extend below the chin, and wrap around the side of the face.

Respirators

PPE also is used to protect workers from hazardous or infectious aerosols, such as *Mycobacterium tuberculosis*. Respirators that filter the air before it is inhaled should be used for respiratory protection.

The most commonly used respirators in health care settings are the N95, N99, or N100 particulate respirators. The device has a sub-micron filter capable of excluding particles that are less than 5 microns in diameter.

Respirators are approved by the CDC's National Institute for Occupational Safety and Health.

Like other PPE, the selection of a respirator type must consider the nature of the exposure and risk involved. For example, N95 particulate respirators might be worn by personnel entering the room of a patient with infectious tuberculosis. However, if a bronchoscopy is performed on the patient, the worker might wear a higher level of respiratory protection, such as a powered air-purifying respirator (PAPR).

Prior to using a respirator, the employer is required to have the worker medically evaluated to determine that it is safe for that person to wear a respirator. The employer is also responsible for ensuring the respirator is fit tested to be sure it is the appropriate size and type, and for training the worker in how and when to use a respirator. The worker is responsible for fit checking the respirator before each use to make sure it has a proper seal.

Proper Use and Disposal

There are four key points to remember about PPE use:

1. First, it should be put on before contact with a patient, generally before entering the room.
2. Once on, PPE should be used carefully to prevent spreading contamination.
3. When work is completed, the PPE should be removed carefully and discarded in the receptacles provided.
4. Finally, one should clean one's hands.

Gowns/Coveralls

The gown or coverall should be donned first. The mask or respirator should be put on next and properly adjusted to fit. The goggles or face shield should be put on next and the gloves are last. The combination of PPE used, and therefore the sequence for putting it on, will be determined by the precautions that need to be taken.

If using a gown, rather than a coverall, there may be some additional process involved when putting it on. To don a gown, first select the appropriate type for the task and the right size. The opening of the gown should be in the back; secure the gown at the neck and waist. If the gown is too small to fully cover the torso, use two gowns. Put on the first gown with the opening in front and the second gown over the first with the opening in the back.

Goggles and Face Shields

If eye protection is needed, either goggles or a face shield should be worn. Position either device over the face and/or eyes and secure to head using the attached ear pieces or head band. Adjust to fit comfortably. Goggles should feel snug but not tight.

The employer is responsible for the establishment and maintenance of a respiratory protection program. When it is not possible to remove the hazard, or while the hazard is being removed, appropriate respirators must be used. An applicable and suitable respirator must be provided to each employee for the hazard involved.

There are many different types and categories of respirators.

Masks

Some masks are fastened with ties, others with elastic. If the mask has ties, place the mask over mouth, nose, and chin. Fit the flexible nose piece to the form of the nose bridge. Then, tie the upper set at the back of the head and the lower set at the base of the neck.

If a mask has elastic head bands, separate the two bands, and hold the mask in one hand and the bands in the other. Place and hold the mask over nose, mouth, and chin, then stretch the bands over the head and secure them comfortably: one band on the upper back of the head, the other below the ears at the base of the neck.

Adjust the mask so it fits securely on the head and fits snugly around the face so there are no gaps. Avoid adjusting or touching it during use.

Respirators

The technique for donning a particulate respirator, such as an N95, N99, or N100, is similar to putting on a pre-formed mask with elastic head bands. Key differences, however, are 1) the need to first select a respirator for which one has been fit tested; and 2) fit checking the device before entering an area where there may be airborne infectious disease. Be sure to follow the manufacturer's instructions for donning the device. In some instances, the manufacturer's instructions may differ slightly from the presentation in this text.

If an elastomeric or powered air purifying respirator, or PAPR, is required, instructions on how to use such a device will be provided at the job site. They are outside the scope of this text.

Gloves

The last item of PPE to be donned is a pair of gloves. Be sure to select the type of glove needed for the task in the size that best fits you. Insert each hand into the appropriate glove and adjust as needed for comfort and dexterity. If wearing an isolation gown, tuck the gown cuffs securely under each glove. This provides a continuous barrier protection for the skin.

In addition to wearing PPE, use safe work practices. Avoid contaminating yourself by keeping hands away from face and not touching or adjusting PPE. Remove gloves if they become torn and perform hand hygiene before putting on a new pair of gloves. Avoid spreading contamination by limiting surfaces and items touched with contaminated gloves.

Disposal

To remove PPE safely, it is necessary to understand what portions of the PPE are considered "clean" and what are "contaminated." In general, the outside front and sleeves of the isolation gown and outside front of the goggles, mask, respirator and face shield are considered "contaminated," regardless of whether there is visible soil. Also, the outsides of the gloves are contaminated.

The areas that are considered "clean" are the parts that will be touched when removing PPE. These include inside the gloves; inside and back of the gown, including the ties; and the ties, elastic, or ear pieces of the mask, goggles and face shield.

The sequence for removing PPE is intended to limit opportunities for self-contamination. The gloves are considered the most contaminated pieces of PPE and are therefore removed first. The face shield or goggles are next because they are more cumbersome and would interfere with removal of other PPE. The gown is third in the sequence, followed by the mask or respirator.

The location for removing PPE will depend on the amount and type of PPE worn and the category of isolation a patient is on, if applicable. If only gloves are worn as PPE, it is safe to remove and discard them in the patient room. When a gown or full PPE is worn, PPE should be removed at the doorway or in an anteroom. Respirators should always be removed outside the patient room, after the door is closed. Hand hygiene should be performed after all PPE is removed.

The following summarizes the sequence for removing and discarding PPE:

1. Using one gloved hand, grasp the outside of the opposite glove near the wrist. Pull and peel the glove away from the hand. The glove should now be turned inside-out, with the contaminated side now on the inside. Hold the removed glove in the opposite gloved hand.
2. Slide one or two fingers of the ungloved hand under the wrist of the remaining glove. Peel glove off from the inside, creating a bag for both gloves. Discard in waste container.
3. Using ungloved hands, grasp the "clean" ear or head pieces and lift

away from face. If goggle or face shield are reusable, place them in a designated receptacle for subsequent reprocessing. Otherwise, discard them in the waste receptacle.

4. Unfasten the gown ties with the ungloved hands. Slip hands underneath the gown at the neck and shoulder, and peel away from the shoulders. Slip the fingers of one hand under the cuff of the opposite arm. Pull the hand into the sleeve, grasping the gown from inside. Reach across and push the sleeve off the opposite arm. Fold the gown towards the inside and fold or roll into a bundle. (Only the "clean" part of the gown should be visible.) Discard into the waste or linen container, as appropriate.
5. The front of the mask is considered contaminated and should not be touched. Remove by handling only the ties or elastic bands starting with the bottom then top tie or band. Lift the mask or respirator away from the face and discard it into the designated waste receptacle.
6. To remove a particulate respirator, the bottom elastic should be lifted over the head first. Then remove the top elastic. This should be done slowly to prevent the respirator from "snapping" off the face.
7. Hand hygiene is the cornerstone of preventing infection transmission. Perform hand hygiene immediately after removing PPE. If hands become visibly contaminated during PPE removal, wash hands before continuing to remove PPE. Wash hands thoroughly with soap and warm water or, if hands are not visibly contaminated, use an alcohol-based hand rub.

> ### Fact
>
> OSHA requires employers to record and report all work related: fatalities, injuries and illnesses. Recording these work-related items does not mean the employer was at fault, an OSHA rule has been violated or the employee is eligible for benefits.

Hospital Radio Frequency (RF) Systems

There are many types of RF systems used throughout a hospital or patient care facility. Probably the most familiar system is two-way radios used to communicate with rescue personnel on their way to the emergency room. Radio frequency systems that are less familiar are the RF systems spread throughout the hospital or patient care facility itself. Although the term "wireless" is often used to describe an RF system, the use of the term today mainly refers to a wireless local area networking (LAN) system used to connect wireless computing devices to the LAN. Regardless of the use of the terms, RF or wireless, these systems all have the same equipment requirements in common. Any RF system needs radios (transmitters and receivers), feed lines (RF cabling, typically coaxial) and antennas (one for the transmitter and one for the receiver). Antennas may be internal to the radio; therefore, no external feed lines to an antenna are required.

System Overview

There are several types of RF system that could be found in a hospital or patient care facility. The most common types are staff presence systems, pocket pagers, digital signage displays, cordless phone systems, and patient wandering and infant security systems. These systems may be standalone, or they might be integrated into other hospital or patient care facility systems, such as nurse call systems or access control systems.

System Types

Staff Presence Systems. Staff presence systems are specifically designed to track staff location. Staff tracking may be manually or automatically activated. The advantage to knowing a staff member's location at any given time helps to provide efficient deployment of employee assets and ultimately provides for better patient services and reduced cost. Typically, manual presence stations and RF tracking staff follower stations provide notification of the locations of individual staff members to the nurse master station and other intelligent devices within a nurse call system. Staff presence systems may be configured with reporting software. This allows the system log and record staff activity.

With automated staff presence systems, doctors, nurses, and aides wear a small ultrasonic, infrared, or radio frequency (RF) transmitter. These "tracking" devices send a signal to a receiver which is usually located on the ceiling in the patient's room. When the staff member enters or exits the room, the receiver notifies the nurse call system of the action.

Pocket Paging Systems. Industry-standard pocket paging systems have been used in health care facilities for many years. In the past, pagers were used only during life-threatening emergencies for priority calls, such as trauma teams. Today, pocket paging systems can be interconnected with the nurse call system allowing for much greater flexibility. Rather than only the attending physician or trauma teams having access to a pager, the health care facilities may provide each nurse and aide with his or her own pager; then assign specific patients to their care. Patient calls may be automatically routed directly to their "assigned" nurse, as well as the nurse master station. The intent is to provide faster and more efficient service directly to patients. Of course, critical calls still can be directed to trauma teams if necessary.

Pocket paging systems are small portable devices that are clipped to a belt or kept in a pocket. Pocket pagers are usually just a receiver with an alphanumeric display; however, two-way pagers are also in use. Paging signals come from a "service provider" that operates one or several paging transmitters located throughout a municipality, or the paging system may be local to the health care facility. When the paging system is local, a "base station" is used to code the messages and transmit a signal to the appropriate pager. Depending upon the layout of the facility, multiple antenna locations may be required. The paging base station may also be connected to the health care facility's telephone system (PBX) to allow for initiating a page via a telephone.

Patient Wandering Systems. Patient wandering systems aid in the tracking of patients who may become disoriented or confused during their stay in a health care facility. Patient wandering systems help to preserve patient dignity while still allowing freedom of movement. The patient wears a small RF transmitter which sends signals to strategically located receivers. The receivers are usually located near passage doors and control door locking mechanisms. When the patient with a transmitter nears the door, the controller locks it. A keypad may be mounted next to each controlled door so that staff accompanying a patient can enter a security code to temporarily override the lock and allow passage. (Note: some systems do not use a lock; they simply report when a patient with a transmitter has passed through a monitored door.)

Patient wandering systems require patients to wear ankle or wrist bracelets containing a small RF transmitter. Receivers are placed at monitored doors. The receiver may control automatic locks at each door where passage is not allowed. When the receiver detects the presence of a transmitter it automatically locks the door, preventing the patient from passing through. Anyone not wearing a transmitter can pass through the door freely. Where no door locking mechanisms are employed, the receiver sends a signal to a monitored station to alert personnel of the patient's location.

Infant Security Systems. Similar to patient wandering systems are infant security systems. The primary difference between the two is the heightened security response offered by the infant security system. Infant security systems may lock-down elevators to prevent doors from opening, activate an alarm if the infant's tag is removed, and locate and/or track an infant. It usually matches a mother (who is also wearing a tag) to her infant with a device that activates an alarm if the wrong tags are paired.

Infant security systems usually have an RF transmitter wrapped around an infant's ankle that emits a constant signal. Receivers are placed at strategic locations throughout the ward. The receivers send signals to the nurses' station and to special door controllers. They control specific doors and electronic locks which energize when the infant is brought within a close proximity. Staff is provided with security pass codes to override the lock via a keypad if the child must be brought to another part of the facility. When a transmitter energizes a lock or if a monitored door is forced open, the nurse master station goes into alarm, alerting personnel.

Personal Emergency Response System (PERS). Personal emergency response systems work in a similar fashion to a home security system. An individual presses a button that he or she may be wearing (a wireless pendant) or button on a keypad mounted on a wall, which notifies a staff member at a remote monitoring service location. The monitoring service then phones the residence that triggered the call. If the issue cannot be resolved over the phone or the individual does not answer, the proper assistance is then dispatched.

Persons living at an independent living facility may have a PERS consisting of two-way audio devices located in their housing units to communicate with staff. A call for assistance is activated via a wireless pendant that is worn around the patron's neck. Activating a pushbutton on the pendant places a call. The person that initiated the call has his or her location displayed when a call is received at a nurse master station. The attending nurse activates the audio device to communicate with the person so that the proper assistance can then be dispatched.

System Components

Personal Emergency Response Systems for a residence utilize a wireless pendant, wall or tabletop push-buttons, and/or a keypad. These systems may be interconnected with a home security system. A remote dialer and the use of a RJ 31X jack would be required (provides line seizure). In an independent living facility, initiating devices similar to the residential version are used (wireless pendent or fob and/or wall or tabletop pushbuttons). However, these devices send their signals to a receiver attached to the nurse master station. The major difference between the systems is the two-way audio system. This is usually an intercom system with a two-way speaker mounted in the wall. (Note: residential systems can offer two-way audio services.)

Wiring Requirements

Wiring requirements for nurse call and ancillary systems used in health care facilities can be found in *NFPA 70*, Article 517 Health Care Facilities.

A nurse call system shall be connected to the "Critical Branch" of the emergency power system. See *NFPA 70*, 517.33 (A)(5) and Article 517, Part VI—Communications, Signaling, Fire Alarm Systems and Systems Less Than 120 Volts, Sections 517.80 through 517.82.

For nursing homes and independent living facilities, see Article 517.42 (A) through (D).

Summary

Working in a health care facility requires special procedures to protect the residents from the harmful effects of construction debris and to protect the electrical worker from infection present in the facility. Work processes must be adapted to shorter cycles to allow for emergency interruptions. Cleanup must occur more frequently in order not to leave construction debris or equipment in the way during breaks or shift changes. Specialized building materials such as lead or some other shielding in the walls of radiation rooms require specialized handling during repair. Finally, the PPE required in a health care setting is unique and something the electrical worker will need help understanding when and how to use.

Review Questions

1. Type A construction is __?__.
   a. less invasive than other types
   b. less complicated than other types
   c. not done in a health care facility
   d. none of the above

2. Replacing cable in an X-ray therapy room would be considered __?__.
   a. Type D construction
   b. medium risk area work
   c. Class IV work
   d. all of the above

3. The need for proper PPE is determined by __?__.
   a. the employer
   b. the worker
   c. OSHA
   d. the CDC

4. The proper PPE for protection against infection is determined by __?__.
   a. the employer
   b. OSHA
   c. the CDC
   d. none of the above

5. Gloves are generally removed before removing a mask or respirator when removing PPE because __?__.
   a. it is easier to undo the other PPE with gloves removed
   b. hands are less sensitive than the respiratory tract to infection and therefore need less protection
   c. gloves are most likely to be contaminated and should not be placed near the face
   d. none of the above

6. Who is responsible for making sure a respirator fits properly?
   a. The worker who uses it
   b. The employer who distributes it
   c. Both of the above
   d. Neither of the above

7. When installing an electrical receptacle in an existing X-ray room corridor wall, __?__.
   a. check to be sure you have the proper lead shielding material for the socket
   b. wear a specially shielded PPE gown
   c. remove all credit cards and metal objects from your pockets
   d. none of the above

8. Before entering or working in an existing MRI room, __?__.
   a. remove all credit cards from your pockets
   b. remember that the power to the MRI machine is always on
   c. remove any metal objects from your pockets
   d. all of the above

9. A Code __?__ is an emergency signal to health care employees that indicates a hospital patient is in some form of arrest .
   a. blue
   b. red
   c. white
   d. yellow

10. Prior to beginning work, the contractor will perform a(n) __?__ and will plan the work to minimize __?__.
    a. cost analysis/expenses
    b. engineering analysis/disruption
    c. risk assessment/rick
    d. safety assessment/injuries

11. After construction is finished and before the area is turned over to the hospital, most areas will need to be cleaned with __?__ filtered vacuums.
    a. AGUA
    b. CERTA™
    c. HEPA
    d. LARA

12. Prior to beginning work in the emergency room and to protect workers, the contractor needs to request a(n) __?__ permit from the hospital administration.
    a. building
    b. demotion
    c. electrical
    d. infection control

Glossary

To aid the student's ability to comprehend the vast diversity of codes and standards within health care facilities, each definition within the Glossary has been supplemented with the addition of the citation of the origination document and section number within the document that the definition can be found.

Nursing Home. A building or portion of a building used on a 24-hour basis for the housing and nursing care of four or more persons who, because of mental or physical incapacity, might be unable to provide for their own needs and safety without the assistance of another person. [**99**:3.3.129] [**70**:517.2]

Packaged Spa or Hot Tub Equipment Assembly. A factory-fabricated unit consisting of a water-circulating, heating, and control equipment mounted on a common base, intended to operate a spa or hot tub. Equipment can include pumps, air blowers, heaters, lights, controls, sanitizer generators, and so forth. [**70**:680.2]

Packaged Therapeutic Tub or Hydrotherapeutic Tank Equipment Assembly. A factory-fabricated unit consisting of water circulating, heating, and control equipment mounted on a common base, intended to operate a therapeutic tub or hydrotherapeutic tank. Equipment can include pumps, air blowers, heaters, lights, controls, sanitizer generators, and so forth. [**70**:680.2]

Patient Bed Location. The location of a patient sleeping bed, or the bed or procedure table of a critical care area. [**70**:517.2]

Patient Care Area. Any portion of a health care facility wherein patients are intended to be examined or treated. Areas of a health care facility in which patient care is administered are classified as general care areas or critical care areas. The governing body of the facility designates these areas in accordance with the type of patient care anticipated and with the following definitions of the area classification.

> Informational Note: Business offices, corridors, lounges, day rooms, dining rooms, ro similar areas typically are not classified as patient care areas. [**70**:517.2]

General Care Areas. Patient bedrooms, examining rooms, treatment rooms, clinics, and similar areas in which it is intended that the patient will come in contact with ordinary appliances such as a nurse call system, electric beds, examining lamps, telephones, and entertainment devices. [**99**, 2005] [**70**:517.2]

Critical Care Areas. Those special care units, intensive care units, coronary care units, angiography laboratories, cardiac catheterization laboratories, delivery rooms, operating rooms, and similar areas in which patients are intended to be subjected to invasive procedures and connected to line-operated, electromedical devices. [**70**:517.2]

Wet Procedure Locations. Those spaces within patient care areas where a procedure is performed and that are normally subject to wet conditions while patients are present. These include standing fluids on the floor or drenching of the work area, either of which condition is intimate to the patient or staff. Routine housekeeping procedures and incidental spillage of liquids do not define a wet procedure location. [**70**:517.2]

Patient Care Vicinity. In an area in which patients are normally cared for, the patient care vicinity is the space with surfaces likely to be contacted by the patient or an attendant who can touch the patient. Typically in a patient room, this encloses a space within the room not less than 1.8 m (6 ft) beyond the perimeter of the bed in its nominal locations, and extending vertically not less than 2.3 m (7½ ft) above the floor. [**99**:3.3.140] [**70**:517.2]

For instance, the definition of *Patient Bed Location* can be found within NFPA 70: *National Electrical Code* in Section 517.2.

Accessible, Readily (Readily Accessible). Capable of being reached quickly for operation, renewal, or inspections without requiring those to whom ready access is requisite to climb over or remove obstacles or to resort to portable ladders, and so forth. [**70**:100]

Alternate Power Source. One or more generator sets, or battery systems where permitted, intended to provide power during the interruption of the normal electrical services or the public utility electrical service intended to provide power during interruption of service normally provided by the generating facilities on the premises. [**70**:517.2]

Ambulatory Health Care Occupancy. An occupancy used to provide services or treatment simultaneously to four or more patients that provides, on an outpatient basis, one or more of the following:

(1) Treatment for patients that renders the patients incapable of taking action for self-preservation under emergency conditions without assistance of others.

(2) Anesthesia that renders the patients incapable of taking action for self-preservation under emergency conditions without the assistance of others.

(3) Emergency or urgent care for patients who, due to the nature of their injury or illness, are incapable of taking action for self-preservation under emergency conditions without assistance of others. [**101**:3.3.178.1] [**70**:517.2]

Anesthetizing Location. Any area of a facility that has been designated to be used for the administration of any flammable or nonflammable inhalation anesthetic agent in the course of examination or treatment, including the use of such agents for relative analgesia. [**70**:517.2]

Bonded (Bonding). Connected to establish electrical continuity and conductivity. [**70**:100]

Bonding Conductor or Jumper. A reliable conductor to ensure the required electrical conductivity between metal parts required to be electrically connected. [**70**:100]

Bonding Jumper, Equipment. The connection between two or more portions of the equipment grounding conductor. [**70**:100]

Bonding Jumper, Main. The connection between the grounded circuit conductor and the equipment grounding conductor at the service. [**70**:100]

Branch Circuit. The circuit conductors between the final overcurrent device protecting the circuit and the outlet(s). [**70**:100]

Branch Circuit, Appliance. A branch circuit that supplies energy to one or more outlets to which appliances are to be connected and that has no permanently connected luminaires that are not a part of an appliance. [**70**:100]

Branch Circuit, General Purpose. A branch circuit that supplies two or more receptacles or outlets for lighting and appliances. [**70**:100]

Branch Circuit, Individual. A branch circuit that supplies only one utilization equipment. [**70**:100]

Branch Circuit, Multiwire. A branch circuit that consists of two or more ungrounded conductors that have a voltage between them, and a grounded conductor that has equal voltage between it and each ungrounded conductor of the circuit and that is connected to the neutral or grounded conductor of the system. [**70**:100]

> Revised for the 2011 *NEC*, 517.18(A) and 19(A) both prohibit the use of multiwire branch circuits from serving patient bed locations in both general care and critical care areas of all health care facilities. This prohibition includes multiwire branch circuits supplied from the critical branch and the normal branch. Multiwire branch circuits are permitted to be used for other circuits and locations.
>
> 210.4(B) requires that ungrounded conductors of multiwire branch circuits be provided with a means to be simultaneously disconnected at the point where the branch circuit originates. This is generally accomplished using multi-pole circuit breakers or identified handle ties. 210.4(D) also addresses multiwire branch circuits from the standpoint that they must be grouped with their associated grounded (neutral) conductor at least once as the circuit conductors enter the enclosure where the branch circuit originates.
>
> The reason the *Code* prohibits multiwire branch circuits in patient bed locations is that a short circuit or overload on one circuit could effectively cause two other circuits of the multiwire branch circuit to automatically disconnect leaving all three circuits of the multiwire branch circuit without power. Since patient care receptacles may be used for life support equipment, this interruption could endanger the life of a patient.

Centers for Disease Control (CDC). OSHA specifies the circumstances under which PPE is required. From a medical point of view, the actual PPE, how to use it, and when to use it are covered by the Centers for Disease Control.

Cord-and-Plug-Connected Lighting Assembly. A lighting assembly consisting of a luminaire intended for installation in the wall of a spa, hot tub, or storable pool, and a cord-and-plug-connected transformer. [**70**:680.2]

Critical Branch. A subsystem of the emergency system consisting of feeders and branch circuits supplying energy to task illumination, special power circuits, and selected receptacles serving areas and functions related to patient care and that are connected to alternate power sources by one or more transfer switches during interruption of normal power source. [**99**:3.3.26] [**70**:517.2]

Dry-Niche Luminaire. A luminaire intended for installation in the floor or wall of a pool, spa, or fountain in a niche that is sealed against the entry of water. [**70**:680.2]

Electrical Life-Support Equipment. Electrically powered equipment whose continuous operation is necessary to maintain a patient's life. [**99**:3.3.37] [**70**:517.2]

Emergency Power Supply (EPS). The source of the required capacity and quality for an emergency power supply system. [**110**:3.3.4]

Emergency Power Supply System (EPSS). A complete functioning (EPS) system coupled to a system of conductors, disconnecting means and overcurrent protective devices, transfer switches, and all control, supervisory, and support devices up to and including the load terminals of the transfer equipment needed for the system to operate as a safe and reliable source of electric power. [**110**:3.3.5]

Emergency System. A system of circuits and equipment intended to supply alternate power to a limited number of prescribed functions vital to the protection of life and safety. [**99**:3.3.41] [**70**:680.2]

Equipment, Fixed. Equipment that is fastened or otherwise secured at a specific location. [**70**:680.2]

Equipment, Portable. Equipment that is actually moved or can easily be moved from one place to another in normal use. [**70**:680.2]

Equipment, Stationary. Equipment that is not easily moved from one place to another in normal use. [**70**:680.2]

Equipment System. A system of circuits and equipment arranged for delayed, automatic, or manual connection to the alternate power source and that serves primarily 3-phase power equipment. [**70**:517.2]

Essential Electrical System. A system comprised of alternate sources of power and all connected distribution systems and ancillary equipment, designed to ensure continuity of electrical power to designated areas and functions of a health care facility during disruption of normal power sources, and also to minimize disruption within the internal wiring system. [**99**:3.3.44] [**70**:517.2]

Explosionproof Equipment. Equipment enclosed in a case that is capable of withstanding an explosion of a specified gas or vapor that may occur within it and of preventing the ignition of a specified gas or vapor surround the enclosure by sparks, flashes, or explosion of the gas or vapor within, and that operates at such an external temperature that a surrounding flammable atmosphere will not be ignited thereby. [**70**:100]

Exposed Conductive Surfaces. Those surfaces that are capable of carrying electric current and that are unprotected, unenclosed, or unguarded, permitting personal contact. Paint, anodizing, and similar coatings are not considered suitable insulation, unless they are listed for such use. [**70**:517.2]

Fault Hazard Current. See *Hazard Current.*

Feeder. All circuit conductors between the service equipment, the source of a separately derived system, or other power supply source and the final branch-circuit overcurrent device. [**70**:100]

Flammable Anesthetics. Gases or vapors, such as fluroxene, cyclopropane, divinyl ether, ethyl chloride, ethyl ether, and ethylene, which may form flammable or explosive mixtures with air, oxygen, or reducing gases such as nitrous oxide. [**70**:517.2]

Flammable Anesthetizing location. Any area of the facility that has been designated to be used for the administration of any flammable inhalation anesthetic agents in the normal course of examination or treatment. [**70**:517.2]

Forming Shell. A structure designed to support a wet-niche luminaire assembly and intended for mounting in a pool or fountain structure. [**70**:680.2]

Grounded (Grounding). Connected (connecting) to ground or to a conductive body that extends the ground connection. [**70**:100]

Ground Fault. An unintentional, electrically conducting connection between an ungrounded conductor of an electrical circuit and the normally non-current-carrying conductors, metallic enclosures, metallic raceways, metallic equipment, or earth. [**70**:100]

Ground-Fault Circuit Interrupter (GFCI). A device intended for the protection of personnel that functions to de-energize a circuit or portion thereof within an established period of time when a current to round exceeds the values established for a Class A device.

Informational Note: Class A ground-fault circuit interrupters trip when the current to ground is 6mA or higher and do not trip when the current to ground is less than 4 mA. For further information, see UL 943, Standard for Ground-Fault Circuit Interrupters. [**70**:100]

Ground-Fault Protection of Equipment. A system intended to provide protection of equipment from damaging line-to-ground fault currents by operating to cause a disconnecting means to open all ungrounded conductors of the faulted circuit. This protection is provided at current levels less than those required to protect conductors from damage through the operation of a supply circuit overcurrent device. [**70**:100]

Grounding Conductor, Equipment (EGC). The conductive path(s) installed to connect normally non-current-carrying metal parts of equipment together and to the system grounded conductor or to the grounding electrode conductor, or both.

Informational Note No. 1 it is recognized that the equipment grounding conductor also performs bonding.

Informational Note No. 2 See *NEC* 250.118 for a list of acceptable equipment grounding conductors. [**70**:100]

Hazard Current. For a given set of connections in an isolated power system, the total current that would flow through a low impedance if it were connected between either isolated conductor and ground. [**70**:517.2]

Fault Hazard Current. The hazard current of a given isolated system with all devices connected except the line isolation monitor. [**70**:517.2]

Monitor Hazard Current. The hazard current of the line isolation monitor alone. [**70**:517.2]

Total Hazard Current. The hazard current of a given isolated system with all devices, including the line isolation monitor, connected. [**70**:517.2]

Health Care Facilities. Buildings or portions of buildings in which medical, dental, psychiatric, nursing, obstetrical, or surgical care are provided. Health care facilities include, but are not limited to, hospitals, nursing homes, limited care facilities, clinics, medical and dental offices, and ambulatory care centers, whether permanent or movable. [**70**:517.2]

Hospital. A building or portion thereof used on a 24-hour basis for the medical, psychiatric, obstetrical, or surgical care of four or more inpatients. [**101**:3.3.124] [**70**:517.2]

Hydromassage Bathtub. A permanently installed bathtub equipped with a recirculating piping system, pump, and associated equipment. It is designed so it can accept, circulate, and discharge water upon each use. [**70**:680.2]

Isolated Power System. A system comprising an isolating transformer or its equivalent, a line isolation monitor, and its ungrounded circuit conductors. [**70**:517.2]

Isolation Transformer. A transformer of the multiple-winding type, with the primary and secondary windings physically separated, which inductively couples its secondary winding(s) to circuit conductors connected to its primary winding(s). [**70**:517.2]

Life Safety Branch. A subsystem of the emergency system consisting of feeders and branch circuits, meeting the requirements of Article 700 and intended to provide adequate power needs to ensure safety to patients and personnel, and that are automatically connected to alternate power sources during interruption of the normal power source. [**99**:3.3.96] [**70**:517.2]

Limited Care Facility. A building or portion thereof used on a 24-hour basis for the housing of four or more persons who are incapable of self-preservation because of age; physical limitation due to accident or illness; or limitations such as mental retardation/developmental disability, mental illness, or chemical dependency. [**99**:3.3.97] [**70**:517.2]

Line Isolation Monitor. A test instrument designed to continually check the balanced and unbalanced impedance from each line of an isolated circuit to ground and equipped with a built-in test circuit to exercise the alarm without adding to the leakage current hazard. [**70**:517.2]

Low Voltage Contact Limit. A voltage not exceeding the following values:
(1) 15 volt (RMS) for sinusoidal ac
(2) 21.2 volts peak for nonsinusoidal ac
(3) 30 volts for continuous dc
(4) 12.4 volts peak for dc that is interrupted at the rate of 10 to 200 Hz. [**70**:680.2]

Maximum Water Level. The highest level that water can reach before it spills out. [**70**:680.2]

Monitor Hazard Current. See *Hazard Current.*

No-Niche Luminaire. A luminaire intended for installation above or below the water without a niche. [**70**:680.2]

Nursing Home. A building or portion of a building used on a 24-hour basis for the housing and nursing care of four or more persons who, because of mental or physical incapacity, might be unable to provide for their own needs and safety without the assistance of another person. [**99**:3.3.129] [**70**:517.2]

Packaged Spa or Hot Tub Equipment Assembly. A factory-fabricated unit consisting of a water-circulating, heating, and control equipment mounted on a common base, intended to operate a spa or hot tub. Equipment can include pumps, air blowers, heaters, lights, controls, sanitizer generators, and so forth. [**70**:680.2]

Packaged Therapeutic Tub or Hydrotherapeutic Tank Equipment Assembly. A factory-fabricated unit consisting of water circulating, heating, and control equipment mounted on a common base, intended to operate a therapeutic tub or hydrotherapeutic tank. Equipment can include pumps, air blowers, heaters, lights, controls, sanitizer generators, and so forth. [**70**:680.2]

Patient Bed Location. The location of a patient sleeping bed, or the bed or procedure table of a critical care area. [**70**:517.2]

Patient Care Area. Any portion of a health care facility wherein patients are intended to be examined or treated. Areas of a health care facility in which patient care is administered are classified as general care areas or critical care areas. The governing body of the facility designates these areas in accordance with the type of patient care anticipated and with the following definitions of the area classification.

> Informational Note: Business offices, corridors, lounges, day rooms, dining rooms, ro similar areas typically are not classified as patient care areas. [**70**:517.2]

General Care Areas. Patient bedrooms, examining rooms, treatment rooms, clinics, and similar areas in which it is intended that the patient will come in contact with ordinary appliances such as a nurse call system, electric beds, examining lamps, telephones, and entertainment devices. [**99**, 2005] [**70**:517.2]

Critical Care Areas. Those special care units, intensive care units, coronary care units, angiography laboratories, cardiac catheterization laboratories, delivery rooms, operating rooms, and similar areas in which patients are intended to be subjected to invasive procedures and connected to line-operated, electromedical devices. [**70**:517.2]

Wet Procedure Locations. Those spaces within patient care areas where a procedure is performed and that are normally subject to wet conditions while patients are present. These include standing fluids on the floor or drenching of the work area, either of which condition is intimate to the patient or staff. Routine housekeeping procedures and incidental spillage of liquids do not define a wet procedure location. [**70**:517.2]

Patient Care Vicinity. In an area in which patients are normally cared for, the patient care vicinity is the space with surfaces likely to be contacted by the patient or an attendant who can touch the patient. Typically in a patient room, this encloses a space within the room not less than 1.8 m (6 ft) beyond the perimeter of the bed in its nominal locations, and extending vertically not less than 2.3 m (7^1/$_2$ ft) above the floor. [**99**:3.3.140] [**70**:517.2]

Patient Equipment Grounding Point. A jack or terminal that serves as the collection point for redundant grounding of electric appliances serving a patient care vicinity or for grounding other items in order to eliminate electromagnetic interference problems. [**99**:3.3.141] [**70**:517.2]

Permanently Installed Swimming, Wading, Immersion, and Therapeutic Pools. Those that are constructed in the ground or partially in the ground, and all others capable of holding water in a depth greater than 42 inches (1.0 m), and all pools installed inside of a building, regardless of water depth, whether or not served by electrical circuits of any nature. [**70**:680.2]

Pool. Manufactured or field-constructed equipment designed to contain water on a permanent or semipermanent basis and used for swimming, wading, immersion, or therapeutic purposes. [**70**:680.2]

Pool Cover, Electrically Operated. Motor-driven equipment designed to cover and uncover the water surface of a pool by means of a flexible sheet or rigid frame. [**70**:680.2]

Reference Grounding Point. The ground bus of the panelboard or isolated power system panel supplying the patient care area. [**70**:517.2]

Relative analgesia. A state of sedation and partial block of pain perception produced in a patient by the inhalation of concentrations of nitrous oxide insufficient to produce loss of consciousness (conscious sedation). [**70**:517.2]

Selected Receptacles. A minimum number of electrical receptacles to accommodate appliances ordinarily required for local tasks or likely to be used in patient care emergencies. [**70**:517.2]

Self-Contained Spa or Hot Tub. Factory-fabricated unit consisting of a spa or hot tub vessel with all water-circulating, heating, and control equipment integral to the unit. Equipment can include pumps, air blowers, heaters, lights, controls, sanitizer generators, and so forth. [**70**:680.2]

Self-Contained Therapeutic Tubs or Hydrotherapeutic Tanks. A factory-fabricated unit consisting of a therapeutic tub or hydrotherapeutic tank with all water-circulating, heating, and control equipment integral to the unit. Equipment may include pumps, air blowers, heaters, light controls, sanitizer generators, and so forth. [**70**:680.2]

Spa or Hot Tub. A hydromassage pool, or tub for recreational or therapeutic use, not located in health care facilities, designed for immersion of users, and usually having a filter, heater, and motor-driven blower. It may be installed indoors or outdoors, on the ground or supporting structure, or in the ground or supporting structure. Generally, a spa or hot tub is not designed or intended to have its contents drained or discharged after each use. [**70**:680.2]

Storable Swimming, Wading, or Immersion Pool. Those that are constructed on or above the ground and are capable of holding water to a maximum depth of 1.0 m (42 in.), or a pool with nonmetallic, molded polymeric walls or inflatable fabric walls regardless of dimension. [**70**:680.2]

Switch.

Automatic Transfer Switch. Self-acting equipment for transferring one or more load conductor connections from one power source to another. [**110**:3.3.9.1]

Bypass-Isolation Switch. A manually operated device used in conjunction with an automatic transfer switch to provide a means of directly connecting load conductors to a power source and disconnecting the automatic transfer switch. [**110**:3.3.9.2]

Nonautomatic Transfer Switch. A device, operated by direct manpower or electrical remote manual control, for transferring one or more load conductor connections from one power source to another. [**110**:3.3.9.3]

Task Illumination. Provision for the minimum lighting required to carry out necessary tasks in the described areas, including safe access to supplies and equipment, and access to exits. [**70**:517.2]

Through-Wall Lighting Assembly. A lighting assembly intended for installation above grade, on or through the wall of a pool, consisting of two interconnected groups of components separated by the pool wall. [**70**:680.2]

Total Hazard Current. See *Hazard Current.*

Ungrounded. Not connected to ground or a conductive body that extends the ground connection. [**70**:100]

Wet-Niche Luminaire. A luminaire intended for installation in a forming shell mounted in a pool or fountain structure where the luminaire will be completely surrounded by water. [**70**:680.2]

X-Ray Installations, Long-Time Rating. A rating based on an operating interval of 5 minutes or longer. [**70**:517.2]

X-Ray Installations, Mobile. X-ray equipment mounted on a permanent base with wheels, casters, or a combination of both to facilitate moving the equipment while completely assembled. [**70**:517.2]

X-Ray Installations, Momentary Rating. A rating based on an operating interval that does not exceed 5 seconds. [**70**:517.2]

X-Ray Installations, Portable. X-ray equipment designed to be hand carried. [**70**:517.2]

X-Ray Installations, Transportable. X-ray equipment to be conveyed by a vehicle or that is readily disassembled for transport by a vehicle. [**70**:517.2]

Notes

Notes

Notes

Notes

Notes

Notes